DOCUMENTS ON MODERN HISTORY OF ECONOMIC THOUGHT

RESEARCH IN THE HISTORY OF ECONOMIC THOUGHT AND METHODOLOGY

Series Editors: Warren J. Samuels and Jeff E. Biddle

RESEARCH IN THE HISTORY OF ECONOMIC THOUGHT AND
METHODOLOGY VOLUME 21-C

DOCUMENTS ON MODERN HISTORY OF ECONOMIC THOUGHT

EDITED BY

WARREN J. SAMUELS

*Department of Economics, Michigan State University,
East Lansing, MI 48824, USA*

2003

JAI
An Imprint of Elsevier Science

Amsterdam – Boston – London – New York – Oxford – Paris
San Diego – San Francisco – Singapore – Sydney – Tokyo

ELSEVIER SCIENCE Ltd
The Boulevard, Langford Lane
Kidlington, Oxford OX5 1GB, UK

First edition 2003

Library of Congress Cataloging in Publication Data
A catalog record from the Library of Congress has been applied for.

British Library Cataloguing in Publication Data
A catalogue record from the British Library has been applied for.

ISBN: 0-7623-0998-9
ISSN: 0743-4154 (Series)

∞ The paper used in this publication meets the requirements of ANSI/NISO Z39.48-1992 (Permanence of Paper).
Printed in The Netherlands.

CONTENTS

MISCELLANEOUS MATERIALS

LIST OF CONTRIBUTORS

Alberto Baccini

Dipartimento di Scienze Economiche, Universita a Degli Studi di Firenze, Firenze, Italy

Warren J. Samuels

Department of Economics Michigan State University, East Lansing, MI, USA

LECTURE NOTES BY
VICTOR E. SMITH

VICTOR E. SMITH'S NOTES ON WILLIAM JAFFÉ'S LECTURES ON GENERAL EQUILIBRIUM, 1938–1939

Edited by Warren J. Samuels

William Jaffé's graduate course at Northwestern University on general equilibrium represented a combination of his interests: the history of economic thought, the work of Jaffé himself on Walras, and general equilibrium theory itself. Victor E. Smith, whose notes from others of Jaffé's courses have been published in earlier volumes in this series, took the course in the Fall semester of 1938–1939.

For personal information on Jaffé and Smith, see archival supplement Volume 6 of this series.

The notes have been edited in the same manner as have other notes by Smith. Presented here substantially as they were recorded by Smith, only minor typographical corrections and stylistic changes have been made. Journal citations have been corrected and completed. The underlining is reproduced from the original notes. Abbreviations have been completed; where any question in this or some other regard arises, square brackets are used. Not all paragraph breaks are clear. Neither diagrams nor other mathematics are reproduced, though descriptive summaries of the former are given in square brackets: it is the topics covered and ideas that are important here. The objective is to indicate what was covered and, to some extent, how it was covered. Most omissions of mathematics are not noted. All notes are the editor's.

Smith numbered his pages by date and page for each date; thus $10/4_{10}$ refers to page ten of the notes for October 4th. The course met once a week, on

Documents on Modern History of Economic Thought, Volume 21-C, pages 3–56.
© 2003 Published by Elsevier Science Ltd.
ISBN: 0-7623-0998-9

Tuesdays, for a two-hour period, so much was covered each session. Smith also indicated the date on which he read each assigned reading; these have been omitted.

In addition to the lecture notes, for the most part written on both sides of $8^{1}/_{2} \times 11$ inch paper, Smith prepared a summary of General Equilibrium on nine blue book pages, c. 7×8 inches. This summary is placed where it was found, at the end of the lecture notes. It also includes, on its last page, a summary of the summary.

Partial equilibrium analysis deals with one product or one factor, i.e. one market, at a time. General equilibrium analysis deals with the interdependence of equilibrium prices and quantities across input and output markets. The focus is on equilibrium, especially the conditions of existence, uniqueness if possible, and stability of equilibrium. Typically the domain has been that of a competitive market economy but it has been extended to non-competitive, e.g. monopolistic, markets. Also, whereas traditional fiscal theory analyzed the expenditure and revenue sides of government budgets separately, general equilibrium theory analyzes their interdependence.

Among the respects in which the notes are interesting are the following:

(1) If one interprets general equilibrium as interdependence, then the scope of general-equilibrium theory is ambiguous. If narrow, it refers, as is generally the case here, to interdependence between commodities and between markets. If very broad, the history of economics can be read quite differently, and the meaning of "explanation" changes: a Ricardian corn-linen model is transformed from partial to general equilibrium. Central to this transformation is the cessation of analysis in terms of unilinear cause and effect and the adoption of mutual interdependency, i.e. mutual dependence of all elements in the price mechanism. (Of course, even partial equilibrium analysis often involved reading causation into statements of equilibrium conditions, such as in Clarkian marginal productivity theory. Coupling the equilibrium analysis with different antecedent normative assumptions yields different paradigms of distribution; as the notes say, "Curious sequels to cost of production theory. Marxian socialists saw robbery. Harmonists saw every share prima-facie evidence of a contribution of sacrifice.") Much the same thing is true of models combining cost of production and utility for each of at least two commodities. The story in terms of value theory, as distinct from price theory, becomes both more complicated and more metaphysical as a quest for the absolute and invariant basis of price – as distinct, say, from identifying the valuation process (of the registration and valorization of interests). In retrospect, it is remarkable

that the mathematization of price theory did not end the reading into price theory of causal and other unilinear thinking. But the main point here must be the extent to which Jaffé understood general equilibrium to deal with interdependence between (the demands and supplies of) commodities – something seemingly much narrower than, but actually not all that different from, modern general equilibrium theory.

(2) The notes represent one episode, as it were, in the development of micro-economics from partial- to general-equilibrium analysis. The meaning of general equilibrium was being worked out, complicated by the parallel dichotomy of statics and dynamics. Did general equilibrium automatically signify transition from statics to dynamics, and what were both "general equilibrium" and "dynamic" taken to mean?

(3) Part of the foregoing may represent, if only in part, something different: A felt need by Jaffé, to review certain elements of then-standard price theory. The reader of the notes becomes aware of the high degree of repetition of treatment of various topics. Yet, one wonders how the course, called general equilibrium, was thought by anyone to differ from a "conventional" micro-economic theory course except in going beyond partial equilibrium. But, certainly, the course did train students in both the *techniques* covered therein and the relevant literature.

(4) Although John R. Hicks' *Value and Capital* (New York: Oxford University Press, 1939) is mentioned by a description of content that Smith took to be a title, it is not central to the course. Jaffé's course, as recorded here, is an indicator of the state of "general equilibrium" theory largely prior to Hicks' epochal work. The chief theorist pervading the course is Walras, or Jaffé's interpretation of Walras – and Jaffé eventually became a leading interpreter of Walras.

(5) Perhaps the most fascinating discussion, at least to this writer, deals with the several ways in which economists have attempted, by strategic assumption, to finesse indeterminacy and non-uniqueness – as the neoclassical research protocol of unique determinate optimal equilibrium results was then being worked out.

(6) Also of considerable interest is the way in which establishing the conditions of equilibrium – e.g. marginal utilities proportional to prices – is considered an "explanation" in the context of solving hypothetical puzzle situations under the assumption that individuals seek to maximize satisfaction.

(7) Close attention to comparisons drawn by Jaffé will reveal differences of formulation now largely forgotten but once of interest to theorists. Also of some interest are the perspectives or interpretations the notes record Jaffé

taking on the meaning of a number of points of theory and the significance of different formulations.

(8) Consider the different treatments of income given by Marshall, Schultz, Hicks, Friedman, and others in the construction of their respective demand curves (surveyed in the lectures). Conceptually very different demand curves emerge from different treatments of income. All are models. No theorist, surely not Jaffé in these lectures, approached the problem in a manner other than how to deductively derive a subjectively suitable, rigorously grounded demand curve. There are no clear and conclusive criteria for choosing between different models; and no-one seems to have been interested in the simple question, what are the factors and forces that govern demand in actual economies? Some or many of these factors and forces arise in discussion, but they are not the center of attention.

Economics D10
ECONOMICS OF GENERAL EQUILIBRIUM

Professor William Jaffé,
Notes by Victor E. Smith
1938–1939

September 20

General equilibrium concept as old as the science. Classicists and certain [others] before had vision of it, intuitive and was not this which neo-classicists stressed in England and U.S. Latter's notions of equilibrium mainly of partial.

Equilibrium: state in any economic quantity or relationship such that any deviation sets in action forces which restore it. Abstract completely from unstable "equilibrium." Not an equilibrium.

Simplest form of partial: Given a demand schedule and a supply for a good all units of which identical. Continuous. [Diagram with demand and supply curves, indicating equilibrium price and quantity; any deviation leads to restoration.] Pure theory of exchange. Can think of in theory of production. Annual rate of output, [if exceeded, leads to] a loss, and some drop out. Quantity reduced back to [original equilibrium level].

Have here price and quantity of single commodity viewed entirely apart from rest of economic structure. Logical device of abstraction by ceteris paribus.

One of objects is to examine limitations upon this device. Neo-classical school has gone further than above sketch. Interdependence of certain prices recognized. Joint demand, joint supply; rival demand, rival supply. Not absent from the neo-classical literature, but consideration of them not characteristic.

Léon Walras gives first conscious central formulation in Eléments d'Economie Politique Pure. Succinctly described in Enrico Barone, translated in Grundzüge der theoretische Nationalökonomie.

May compare market to complicated machine such that all parts change position if one does.

Henry Ludwell Moore, Synthetic Economics. Balance of many forces, operative on price, configuration which remains in state of rest. Dates from Cournot and Lausanne school (Walras and Vilfredo Pareto). Walras professor at Lausanne in Switzerland, tho was French. Followed by Pareto at Lausanne. Latter modified and improved. Multiplicity of continental economics. In U.S., Moore, Irving Fisher (Mathematical Investigations in the Theory of Value and Prices), Henry Schultz.

Every problem brings important questions of general equilibrium. What effect of higher price upon demand for given commodity? Are other prices changing too? Does change in this demand change other demands? Does there result a change in incomes?

J. R. Hicks, <u>Mathematical Theory of Value under Free Competition</u> [sic]. Essay in general equilibrium. Some developments following Walras and Pareto. Follows plan of Walras.

More practical question of increased government expenditure on public works. Raw material prices. Relief incomes and purchases. Savings. Investments.[1]

Erich Lundberg, <u>Economic Expansion</u>. Concerned with economic phenomena as function of time. Cyclical fluctuations and trends. Finds partial equilibrium analysis inadequate, even when cumulated. Want to know effect of a number of simultaneous changes. For theory of business cycles and secular trends use general equilibrium. Wassily Leontieff, Interrelations of Prices, Output, Savings, and Investment, <u>Review of Economic Studies</u>, 19: 109–132 (August 1937):

Application of general equilibrium. An empirical study. Admissible method – rational empiricism. (1) Formulation of appropriate theoretical scheme. (2) Gathering and arrangement of statistical material. (3) Empirical application of theoretical scheme to data's analysis.

In deciding as to type of theoretical scheme, have two choices. General vs. partial equilibrium, static vs. dynamic analysis.

Partial dynamic seems to dominate now. Comparison of partial dynamic and general static approaches not to be by <u>a priori</u> means. Only complete knowledge could decide and wouldn't need.

Test of empirical application can't be made until finished.

General gives possibility of handling interrelationships. Partial can't succeed, with intuition only, beyond a few variables.

Henry Schultz, <u>The Theory and Measurement of Demand</u>, uses general.

Socio-ethical problems affect whole system. What of declining population of England?

General equilibrium may be of supreme importance.[2]

Should have French, German, Swedish and Italian languages. Should have lot of mathematics. Will try to interpret in non-mathematical fashion. Those who have it can present and privately discuss papers using it. Have translations in to English from Italian. London School, especially Lerner, Mrs. Hicks, and Lionel Robbins, has done much to make available.

SELECT BIBLIOGRAPHY:[3]

A. Cournot, <u>Mathematical Principles of the Theory of Wealth</u>, 1838.
Léon Walras, <u>Eléments d'Economie Politique Pure</u>.

Irving Fisher, Mathematical Investigations in the Theory of Value and Prices.
Knut Wicksell, Über Wert, Kapital und Rente.
Knut Wicksell, Interest and Prices.
Knut Wicksell, Lectures on Political Economy.
Vilfredo Pareto, Cours d'Economie Politique, 1896, 1897.
Vilfredo Pareto, Manuel d'Economie Politique, 1909.
P. H. Wicksteed, The Common Sense of Political Economy.
Enrico Barone, Grundzüge der theoretische Nationalökonomie, translated from
    the Italian by Staehle.
Joseph Schumpeter, Wesen und Hauptinhalt der Nationalökonomie.
Joseph Schumpeter, Theory of Economic Development.
R. A. Murray, Lécons d'Economie Politique (translated from the Italian; no
    mathematics).
Francois Divisia, Economique Rationnelle.
Atto Kšhne, Die mathematische Schule der Nationalökonomie, Vol. 1.
Henry Ludwell Moore, Synthetic Economics.
Gustav Cassel, The Theory of Social Economy.
J. R. Hicks, Theorie Mathematique de la Valeur.
Erich Lundberg, Economic Expansion.

Read:
Wicksell's Lectures, Vol. I to p. 35.
Moore, Synthetic Economics, to p. 32
Umberto Ricci, Pareto and Pure Economics. Review of Economic Studies, 1:
    3–21 (October 1933) (Translated by Ursula K. Webb [sic: Hicks].
Jaffé, Unpublished Letters and Papers of Léon Walras. Journal of Political
    Economy, 43: 187–207 (April 1935).

Will be short papers almost every week, on matter which we are not expected
to get very well. Prepares us to follow class discussions by having us meet the
problems in advance.

WALRAS' SYSTEM

Fundamentally, five conditions that must be simultaneously realized.

(1) Each individual must so distribute resources over all possible expenditures
    as to obtain maximum satisfaction. It is this that brings into contact with
    Austrian school and marginal utility.

(2) Quantity supplied of each commodity should equal quantity demanded. Condition to be reached through changes in prices. Must coexist with (1). These two underlie the pure theory of exchange.

(3) In pure theory of production: Price = cost of production. Obtained through variations in rate of output. Must coexist with above.

(4) Rate of net yield on capital investments must be the same in all different types. If more in one than in other, a movement of capital between uses.

(5) Quantity of money must be such that price of monetary metal used as money equals price in bullion market. Mint = bullion.

Theory of money, and of cash balances (questionable), incorporated into general equilibrium.

(6) Not a theory of capital formation. Hicks has no money or capital formation, for is out of place in static state.

School of Lausanne does not confine self to simple description of equilibrium, such as above. Is not yet the whole story. Still is partial. If wish to explain a given pattern of prices, must take into account institutional effects. [See notes 1, 2, 3.]

Too restricted. Opponents, however, say to throw over the mathematics. Difficulty is that is insufficient. Need to add to it, not to reject it. In school of Lausanne find proof as well as statement. Indulge in proof of rigorous mathematical character. (May be mistakes, but are exposable.)

Barone says machine works with much friction. Do not have rapid exchange of movement. Slow, and at different rates. Whole tends toward new equilibrium, but upset by new changes before reached. Reality consists in dynamic world. Only insofar as general equilibrium can be carried into analysis of dynamic phenomena of reality are results useful.

Attitude of self-criticism characteristic of general equilibrium economists. Seek ultimate application to reality.[4]

In 1690, Nikolas Bower ["?" above name. This could be Nicholas Oresme, c.1320–1382.] wrote that many lacked true idea of trade because applied thoughts to a particular part of trade and then extended to the whole. Put together parts into a deformed whole.

Classical equilibrium in Smith, Book I, Chapter VII, and in Ricardo, Chapter IV. Both deal with relation of market price to natural price.

Cf. Moore, pp. 22–23.

Important to note that classicists had sketched general equilibrium in bold outline. Either was overlooked or thought too unwieldy later, for particular problems had to deal with.

Marshall gives development of partial. Cournot gives first known use of phrase, general equilibrium. Smith gives price = cost, effective demand = effective supply, interest rates same. Natural price the central price to which all others gravitating. Deviate by accident. This is a "center of repose and continuance."

Marshall saw his extensions beyond his partial equilibrium as the most important of his work. Elaboration of Note 21, interdependence of prices, his life work, he wrote J. B. Clark.

A. Smith not content with partial: In Chapters 8–11 of Book I, examines relationship of profits, wages or rent on different employments. True beginnings of general equilibrium analysis.

Public mourning raises price of black cloth, but not weavers' wages, for not understocked with labor. Wages of workers on the black cloth are raised, for is labor scarcity here. Etc.

For conscious beginnings must go to Cournot, 1838. Chapter XI, "Social Income." Formerly held prices of other commodities and incomes invariable. Up to, was of partial equilibrium. Went on: in reality, an interdependent whole. Increase in demand for A increases its producers' incomes and their demand for other goods, their producers' incomes, and their demand for A. Cournot assumes demand for A elastic so total spent on it declines as price rises. Affects other commodities' demands.

If same annual amount spent, may be no change in distribution of expenditures. Must usually be effects, tho.

Reactions must go on with decreasing amplitude, by general principle of analysis, says Cournot.

Producers of A find incomes varying and thus their demands.

Decreasing amplitude a pure assumption. Doctrine of convergence needs criticism. Cobweb or spiderweb theory: [Diagram illustrating cobweb effect of path,] may or may not converge depending upon relative elasticities. Given time, path by which obtain equilibrium affects result.

General equilibrium only one of great contributions of A. Cournot. Gave hint only. Main reason for not developing was negligibility of indirect effects.

Question: Utility analysis takes prices for granted. Try to attain maximum satisfaction within limits of any given price configuration.

Can have willingness to work for practically no wages, yet no adequate opportunity for employment. An example of discontinuity perhaps. (Assignment on 9/2015.)

September 27
For October 4: Paper (not over four pages) on significance of corn-linen example, pp. 24–5, as criticism of classical theory of value.

Read:

Gustav Cassel, <u>Theory of Social Economy</u>, Chapter I, pp. 1–41.

Wicksteed, to page 43 (from page 35). Note Figure 1, p. 39.

Wicksteed, <u>Common Sense of Political Economy</u>, Book I, Chapters 1 and 2, pp. 13–94.

Have: defined equilibrium, gone from partial to general, described meaning of latter, sketched history of theory and authors, pointed to theoretical, practical and social significance, and looked to method of procedure, bibliography, Walras' theory, and early forms (Smith and Cournot).

Moore's: Each factor seeks and finds maximum net income (satisfaction, says Jaffé), within limits of resources.

Merely another way of saying perfect mobility.

General equilibrium theory assumes, by implication, full employment, at least Walrasian and Paretian schemes. Are enough devices so that could include if wished.

Wicksell, Johan Gustav Knut (1851–1926): Swedish. Emil Sommarin, in <u>Austrian Review</u>, 2: 221–267, best biography [Emil Sommarin, "Das Lebenswerk von Knut Wicksell," *Zeitschrift fur Nationalökonomie*, October 1930]. In English, obituary notice in <u>Economic Journal</u>, September 1926, and <u>Econometrica</u>, Vol. I, no. 2.

In years like these when we are in college or graduate school he didn't study economics, as did not other greats. Mathematics and philosophy; student affairs; wrote musical comedy; joined radical organizations and temperance society. Latter was start of interest in economics. Didn't mean prohibition, but system of rationing by State. Gave importance to ending scourge of drunkenness. Theory that most frequent cause was poverty, and that caused by high birth-rate. Advocated birth control, and reproached by all good of Sweden. Read Malthus by advice of opponents, and then more economics.

Reaction in religious beliefs and practices in youth. Opposed Christian religion in all orthodox forms. Radical in old age as in youth. Imprisoned at 58, and wrote book on population, signing preface with "State Prison." Only economist of repute known to have gone to prison.

Spontaneous interest in people's welfare. A fighter. Wanted a human society free from burden of poverty and its evils. Felt imperative to have small population, much less than had by countries at time. Determined optimum population, in mathematical and statistical form. Defines it. No family with more than two or three children. In 1887 wrote that children being born would find U.S. immigration barriers when came to emigrate.

Scientist at heart. Not blinded from seeing fundamental relationships. At age of 35, traveled and studied for five years, supported by others, studying principally early classics.

First important work in 1893 at age 43, <u>Über Wert, Kapital und Rente</u>. Interest in value from his optimum population – high level of welfare – satisfaction of wants – problems of evaluation. In 1895, only, got doctorate. To become professor, had to study law for four years, for on law faculty. Assistant professor in 1900 at Lund, and full professor in 1905. Retired in 1916. Married Anna Bugge, who became one of Sweden's most prominent women lawyers. Son professor of mathematical statistics at Lund.

Writings appropriate for us, for stresses mutual dependence of all elements in price mechanism. Genius shows by ease of developing profound ideas.

Introductory chapter in four parts: Nature of econ[omics], main divisions, subdivisions of theoretical treatment, and nature of assumptions. Begins historically (showing influence of Schmoller), as did Cournot.

Nationalökonomie – national housekeeping.

Econ[omics]: Doctrine of economic phenomena in interrelations seen as a whole.

Economic phenomena: Systematic endeavor seeking greatest result for given means, or given result for least possible means.

Endeavors such as these involve exchange, and individual self-interest often benefits all of society. Often not. Latter important for itself and as principle of division of subject. After Walras, theoretical, practical, and social econ[omics].

First has aim to determine necessary connections between economic phenomena, which call laws. Necessity is one of logic rather than nature.

Complicated whole can't be grasped by mind, so break into parts by simplifying assumptions. Insofar as partake of reality, necessary connections are also natural.

Practical aims to examine working of these laws in concrete situations, taking existing private property, etc. and see if helps society as whole, future, too.

Social to examine institutional frame and to determine which would give greatest social gain.

Walras uses necessary, expedient, and just.

Wicksell acknowledges Walras tie. Fail to distinguish these problems, nowadays. <u>Very</u> important. Can't determine justice from necessity or expediency.

Sommarin cites question in Denmark for doctorate: What advantages and disadvantages from stabilization of value of money?

Theoretically, ambiguous. If mean purchasing power, idea for smooth working of mechanism.

Applied, difficult job. Disturbances of wars and truces (Jaffé, used to say "post-war"). Unstable economic mechanism and too much to expect any part stable. Private interests can profit from fluctuations, and may benefit whole nation.

Social: Must take wider than national and immediate view. Monetary unions show international cooperation for national stabilization.

Want of clear distinction that confuses discussion of laissez-faire. Appeal to whole science of economics for support. Irrelevant to argue from assumption that must be best. Theory may tell what to expect if had, but doesn't tell if possible. Nothing of historical foundations or ethical-social desirability.

Economic science born out of controversy against mercantilistic controls of late 18th century.

Was possible to define the maximum satisfaction attainable under laissez-faire. Tools applicable to another system, but which maximum is the greater.

Wicksell to deal with pure theory, as we. Can't refrain from remarking that concern with pure theory of whole and interrelations an index of revaluation in attitudes, for laborer as important as landowner.

Walras' ideal a state creating equal opportunity for attaining unequal positions.

Pure theory divided: Rejects traditional chronological division of Mill, production, distribution, exchange, and consumption.

Logical move is Walras': exchange, based on analysis of wants, or desires; production, implicitly involving distribution, for pricing of factors of production; capital formation and credit; money.

Expository and logical significance, tho not chronological.

Wicksell: (1) demand, (a) quantitative (population) and (b) qualitative (value). (2) Production and distribution. (3) Growth of capital. (4) Money and credit.

Cf. with Marshall, who swings to supply at (2) (which is essentially production), then supply and demand together to yield value, then distribution as special case of value. His order of argument unsatisfactory.

Walras an "architect." Nature of fundamental assumptions: Necessary, for complexity. Obvious in experience of institutionalists, who express bewilderment by anger at others.[5]

Must neglect certain features: (1) Irrelevant or minor forces with reference to problem. May take man as self-seeking for some cases but not others. (2) Relevant, if temporary, to simplify intricacy that defies immediate solution. Examples in Wicksell's abstraction from money in discussing value, tho is relevant. International trade likewise – a closed community, or unit.

Production, likewise, as Marshall and Walras. Later introduce production, abstracting from capital formation. Drop as soon as methodologically possible. Not inherent, but methodologically necessary.

Last paragraph of introduction points out abstractions of historical school. Marginal utility and relation to exchange, value, and price.

Object is to explain why one has one price and another a different. What part has marginal utility or significance played in explanations of price configurations?

Controversy between price and value economists, over meaningless issue. We postulate that agree should describe and explain complex of price phenomena as principal task. Price theory the center of economic theory. Only question is how. With or without theory of subjective value, to serve in purely instrumental way as tool of price theory? Similar part to roundabout means of production. Thus says Morgenstern. No fundamental issue, but merely can we have price theory without this? We read Cassel and cf. with Wicksell.

Usual to begin definitions, but objectionable, for arbitrary. Meaning obscure apart from solution of problem.[6]

Definitions irrelevant as well as internally inconsistent are foisted in concepts chapter. Marshall apologizes for traditionally mentioning unproductive labor.

Our problem is subj[ective] value. Needed for price configuration? Mean importance or significance attached to a definite quantity of good (or a little more or less of it" by an individual. Tool for reasoning of differentials is the calculus. Might question, for individual deals with discrete elements, and usual calculus is of continuities.

Exchange defined as choice between various uses of same means of production or finished commodity or various means of achieving same end. Requires only one person for exchange in this sense. Whether a definition is any good, can't be determined by examining the word, but the problem. Production is exchange by this definition. (a) Is logically inconsistent within itself? (b) Is contrary to facts? (c) Is useful?

Exchange only under limiting conditions. Only on definite rate of so much of one thing given or given up for another.

Exchange-value is ratio at which is given for other goods ... [Ellipsis in original.] As many values as are other goods. In n goods, each good has n−1 values. Not counting in terms of itself. For ease, look to ratio between given and majority of others, whether goes up or down relative to the majority.

When says price sometimes same as exchange value, probably thinks of Walras. Subjective value and exchange value: Undeniably related when have rival commodities. If equally efficient, exchange value same. If importance of added unit equal to, greater than, or less than another, exchange values proportioned to subjective values, and to efficiency.

When are two not competing, satisfying different wants, how compare subjective values, and how say one has equal to, less than or greater than subjective value? Fundamental problem in whole history of economics.

Can say that prices proportional to marginal significance in same sense as when competing?

Smith set paradox for one hundred years: Water high in use and low in exchange; diamonds vice versa.

Classicists turned to relative scarcity, given value in use. Scarcity or difficulty of production regulates and controls supply, if utility the demand. Price varies directly as quantity demanded and inversely as quantity supplied.

Schultz paraphrased: Price is proportional to D/S. Inadequacy first demonstrated by Cournot: Never thought that statistics might be lacking, so couldn't deduce applicable consequences. But what principle? Price fall one-half if quantity doubled? Not true, if so stated. Quantity demanded does not mean that actually marketed, for from the principle would get dearer as more marketed, while at low enough price, demand indefinite.

Can't speak of ratio between supply and demand, for equal each other. Ratio always unity. Why equilibrium at one price for one commodity and another for another?

Classicists saw inadequate, and supplemented by cost of production and cost of reproduction. If market price to equal natural price, must have supply = demand and price = cost. Differences in price between two reproducible commodities due to differences in costs. But costs must be definite and independent, not themselves prices. Can't explain unknown term in terms of the unknown, but that is what they do. Only if exactly same conditions of production, can say prices equal. Comparison impossible if proportions differ, unless reduce to common terms.

Example: Table and chairs: twenty hours labor plus twenty hours services of saw and hammer and one year's service of one-quarter A. [acre] land and one dozen chairs=one table [?]. Prices proportional to costs, or if table = three chairs. If takes less labor and more land for table, how cf.? If use common denominator, will be price, and have no explanation. Theory of natural price, of course.

Discuss Ricardo's attempt to get out of difficulty next time.

October 4

In next three weeks, read Wicksteed, Book 1, Chapter III, and Book 2, Chapters I–IV, pp. 95–125.

Cassel, within four weeks, finish Book I, pp. 42–88.

Wicksell, pp. 43–49, for next week.

Walras' "Geometrical Theory of Price," to middle of page 50; next week, to page 54.

Paper interpreting figure 1, p. 39, Wicksell, with own examples and in own language.

Umberto Ricci, "The Modification of the Utility Curve for Money in the Cases of Indivisible Goods and Goods of Increasing Utility," Economica, n.s. 2: 168–197 (May 1935). [In margin: "Merely read once. ?]

Henry Schultz, Interrelations of Demand. Journal of Political Economy, 43: 468–481 (August 1933). Also about 100 pages in his book just out. [In margin: "Merely suggestions," with arrows pointing to Ricci article and second Schultz item.]

Division I of ec[onomics]: necessities.

Division I: Most expedient measures in light of certain necessities and aims, given data of pure theory regarding the necessities.

Division III: Justice, the most just means, etc.

Wicksell's Lectures deal with pure theory alone.

Assumptions chiefly to simplify, so can get comprehensive grasp by successive approximations. Abstract (1) permanently from irrelevant or unimportant matters[7] and (2) temporarily, for methodological reasons.

Demand, production and distribution, capital formation, money and credit, all tied together within system of equilibrium.

Object is to explain price configuration.

Is a given approach useful?

Classical theory had: supply = demand, and price = cost. Defect in fact that costs are selves prices, and still no solution if latter prices not explained. Ricardo's ingenuity: Tried to reduce various items to labor, and posited exchange values similar or in proportion to quantities of labor employed. Reduced all types of labor to standard through relative market prices. Capital chiefly made of advances to aid and support, and considered advance per hour standard labor similar same. Quantity of capital similar or proportional to quantity of labor. Thus could neglect, since is same factor on each side. Land eliminated by going to no-rent margin. If capital for support of labor, proportional to it. Also if for aid of labor, says Jaffé,, for is the machine which is used up which affects price, and this deviates with rate of using. [Smith added in brackets: "Conceivable, but true? If I understand?]

Another assumption: Labor per oz. Au [ounce of gold; Au hereafter gold] invariable. Capital's share same in gold production as in any of all other industries. Labor entering into a good a direct measure of quantity of gold for which it will exchange, or price. General rise in wage level can't affect commodity prices, for ratio between com[modity?] wages and gold wages, and unaffected.

If commodity produced under constant costs, no margins, only on demand side. If increasing, have demand margin (last unit worth buying, given price) and production margin. If have two commodities of increasing costs in closed com[?], relative values depend upon extent to which demand forces margin.

Why Wicksell's corn-linen example? Refute Ricardo in some cases, as step bigger project of showing mutual interdependence? [Equations regarding relative prices of corn and linen and of quantities of labor used in production of corn and linen, respectively, at no-rent margins.] Prices set by world market: Corn low in linen and linen high in corn. Use very little land for corn and need little for linen. Not use inferior land, so little labor per unit corn or per unit flax. Add labor to flax to get linen high in terms of corn. [Equation in which prices of corn and linen are in proportion to respective quantities of labor used in producing single units of linen and of corn. Said to be:] Definition of equilibrium position: Assume here expressing in a third commodity.

If international market changes, so corn goes up in linen, extent and intensify margins. Let quantity of labor in corn double, and in flax double. Quantity in linen doesn't double, for that in manufacturing linen must be considered.

One of main purposes is to show that margin not fixed, but variable, with price, and can't use to explain price. Exchange value governs costs, here. Interesting in re over-simplification of marginal product theory of wages. Marginal product variable, and responds to wages. Whether more or less of employment another question.

Reject this causal nexus, and have left mutual equilibrium.

Also against cost of production theory: Joint supply makes impossible to allocate separate costs. Even tho can vary proportions by varying separable costs, relation is of interdependence, not cause and effect. If anything, operates in other way, from value.

Curious sequels to cost of production theory. Marxian socialists saw robbery. Harmonists saw every share prima-facie evidence of a contribution of sacrifice.

Gossen, Jevons, Menger, Walras, Böhm-Bawerk: Marginal utility and subjective value theory an auxiliary to price theory.

Main problem of explaining price-configuration on market, why two commodities have different prices.

What is meant by "explain"? Plato to Aristotle say to form intelligible universe by finding universals in the particulars. Greeks taught actual particulars the primary facts which exist as related to general classes. [In brackets: symbol for similar and question mark.]

History of value theory is of efforts to fit price phenomena into universal pattern of general validity. Early abandoned utility pursuit because sought general importance or utility. Then sought differences in costs, but non-reproducible commodities excepted, perforce. Attended to generally reproducible. Found cost differences particular phenomena of same sort as prices, and themselves had to go into frame.

Even marginal prod[uctivity] device a blind alley, for margins variable, and have to express both margins and price as part of universal.

Mathematics showed futility of pursuit of ultimate causes. Science seeks particular manifestations of generally logically coherent system.

Can fit subjective value theory into it? Useful only as tool of price theory. Made headway with its use. Cassel maintains it not indispensable.[8]

What is this theory and what its part in price theory?

How about the presumably more objective analysis of indifference curves?

Value in use and value in exchange? Confine attention in each case to manageable quantities. If value in use an upper limit of value in exchange, for buyer will not give more for less, ask also of seller, and find it a lower limit, too. Is value in use = value in exchange? Contrary to experience.

Later shows same thing has different significance for different persons, and also for same person if he has different quantities. Differences in amount of money possessed fundamental in modern society.

As many variables of significance for a slice of bread as individuals, perhaps but only one price, if competition.

Answer, page 30, that degree of utility that has for person at time of exchange is effective – the minimum utility; this degree of utility called the marginal or final utility.

The last need satisfied if get it, is marginal utility. For very small quantities can't distinguish between last taken and first not.

Finally, marginal utilities (degrees of utility) proportional to exchange values, or would exchange.

Price corresponds to least important desire satisfied by small quantity taken at given price.

Price refers to value; and so to exchange, so only if at least two commodities. [In brackets: "Goods"?]

Magnitudes of desire satisfied by last unit taken and last given same for the man and on each side of exchange.

Notion of scale of diminishing utility important. Fuse utility and scarcity, calling relative scarcity, or marginal degree of utility.

Jaffé, feels terminology lacks precision. Would distinguish marginal utility from marginal degree of utility. Latter is du/dx, and former is du/dx for [schedule of] x, following Marshall's final conclusion in note 1, reached in his 4th edition.

Auto at uniform rate of 30 miles per hour. Increment in space in 2 minutes is 1 mile. May hold rate constant, but increment varies with time. Or reduce rate, and increment varies. Rate is ratio, distance/time. Increment (distance) equals rate x time. [Diagram with miles per hour on vertical axis and number of hours on horizontal axis.] Areas are increments. Altitude, by definition, the rate, and base the time. Could find increment (of distance) between 3rd and 5th hours by [multiplying hours by miles per hour].

In subjective value analysis wish to unite: (1) quantity of commodity to be acquired; and (2) anticipated satisfaction at time of acquisition (utility). Assume rate uniform and can express utility in terms of dollars. Buy five years' shoe supplies, each pair representing $5 worth of anticipated satisfaction. [Similar diagram, with rate of anticipated satisfaction per pair of shoes, in dollars, on vertical axis, and number of pairs of shoes on horizontal axis.] Can find rate by dividing $10 [total spending on shoes, in dollars] by $2, if know total anticipated satisfaction [in dollars].

Rate of increase in anticipated satisfaction (utility) not uniform. Dollar's worth not a constant measure.

Utility increases as quantity of good, up to point of satiety. [Diagram with total utility on vertical axis and number of units of good on horizontal, with positively inclined curve with ranges of, first, increasing and, second, decreasing rates. Indication of point of inversion, where ratio of marginal utility to marginal quantity is unity.]

Area under curve has no economic significance. Ordinate measures total utility. Can use areas, but watch significance of axes: [Diagram with utility on vertical axis and quantity on horizontal axis, with curve of increasing, maximum and decreasing marginal utility; indicating that area under curve between two quantities is total utility for that range.] Posit increase for while at increasing rate, and then declining rate, as a vacation in Bermuda, of variable length. On 0y axis, plot additions to total utility as add dX. First unit may have no meaning until add others. [Diagram with utility implicitly on vertical axis and days in Bermuda on horizontal axis.] Utility ascribed to 2nd day not just that received on 2nd day (Monday), but the addition to total utility over what would receive if only had one day.

If an irregular step curve, what of rates and increments? If seek only rate and increment per day, for each day: Rate and increment the same as long as additional discrete units of same size, and equal to unity.

If have total added in 2nd and 3rd days, tho, can't tell rate, for differs. Still worse if go to continuous flow of time, rather than artificially discrete units of days.

When plot discreteness by very small units [result is] similar to continuous curve. If utility received on 5th and 6th day is [arrow to area on decreasing marginal-utility range] and on 21st to 23rd days [arrow pointing to area further down curve]. Rate constantly changing, and how to find at given moment of time?

If an auto, take as short a distance as possible on either side of point, and take some sort of average. [Diagram with "rate (distance/time)" and distance on the axes, with positively inclined curve labeled "if gathering speed."]

Ordinary utility curves measure rate of change of utility with respect to quantity, on 0y axis. (du/dx or du/dq) Area under such a curve represents total quantity of utility acquired.

Definition of equilibrium, taking into account price and utility, is that prices proportional to rates of change of utility, and final increments of utility equal. [In brackets: "If units such that prices equal."]

[Two diagrams, with du/dq and qa on axes, with differently shaped negatively inclined curves.] Total utility [to a level of qa] [is area under curve to that point.] Give now a rate of exchange, [In first diagram,] pa,b = 2 (2 units b = 1 of a). [In second diagram, Gives up 2 of b for 1 of a, and worthwhile if added total utility of a greater than lost total utility of b.

Continues until increase of utility = decrease of utility, permits being infinitely divisible. If goes any further, loses on the exchange. Equilibrium point is where marginal utilities (increments) are equal. Marginal degrees of utility differ, however.

[Mathematics showing foregoing, extended to third commodity.]

Prices proportional to marginal <u>degrees</u> of utility, but marginal utilities themselves are equal. [In brackets: "See prefatory note to paper for October 11."]

October 11
Cassel, pp. 88–131.
Wicksteed, pp. 49–63.
Wicksteed, pp. 401–73.
Walras, pp. 50–54.

Paper on Walras' "Geometrical Theory of the Determination of Prices," to page 50, proving geometrically that [equal ratios of marginal utility to price for series of goods exist] in state of maximum satisfaction.

At bottom of page 71 of Wicksteed, discussion of competing and completing goods. Dip into Henry Schultz, Interrelations of Demand, <u>Journal of Political Economy</u>, 41: 468–512 (August 1933), Secs. (a) and (b), pp. 468–481. [In margin: "Once for 'dipping.'"]

[The next four pages of notes are on the back of Smith's paper, "Corn, Linen, and the Classical Theory of Value," with comments by Jaffé. Jaffé, commented, "Good but diffuse. You might have expanded the proposition at the bottom of page 3." Smith marked "?" after the first sentence of the comment. The proposition referred to is: "The applicability of Wicksell's criticism [of J. S. Mill] is increased by any widening of the class in which costs change with a shifting of the margin due to demand or values, since this tends to become the general case and constant costs the special case." Alongside Jaffé, wrote "Good!" Elsewhere Jaffé, commented about the statement that "price equal to cost" that

"This is simply a definition of one aspect of the equilibrium function."]

Last time tried to give indispensable tool concepts for understanding of theory of maximum satisfaction, from standpoint of utility analysis. Important, for maximum satisfaction a pillar of general equilibrium. No equilibrium if not the maximum satisfaction within limits of resources.

[Diagram with distance and time on axes, with vector gradient bisecting quadrant.] Straight line indicate our car going at a constant rate. Increments of distance [per unit of time] equal. [Equations relating to rate, total distance, total time. New diagram with rate and time on axes.] Area [under constant-rate line] is distance, and varies proportionately with the ordinates of distance function, and with time.

Take car accelerating [...] and seek speed when passing [particular point]. [...] Know, if accelerating, that a rate at each point is conceivable, and may be more closely approximated the finer the measuring. [Diagram with distance and time on axes, with positively inclined curve indicating acceleration.] Increments of distance [per unit of time] are growing as time passes. Rate per hour at end of hour = change in distance/unity.

[Further discussion with diagram on acceleration; then deceleration – "Increments of distance declining" – and comparisons with constant rates of speed.]

Gives some idea of differences between and relationships among rates and increments.

[Diagram with distance and time on axes, with positively inclined curve having accelerating range, range of constant increments, and decelerating range.]

Important for Jevons' final degree of utility: rate of change of utility as function of quantity.

[Diagram with distance and time on axes, with positively inclined curve. Indicates a rate for a period, dy/dx.] Is an average rate for dx, and may be non-representative.

Both Jevons and Walras considered utility a function of quantity of the given commodity alone.

Interrelations of demand extremely complex, and inaccuracies involved in neglecting not large. [Diagram with utility and quantity on axes, with curve first increasing at increasing rate and then decreasing rate.] Convex upward, save perhaps for first portion. Total utility increases, at decreasing rate, to point of satiation, so far away that can neglect. Increments decreasing progressively.

[Diagram with dy/dx and quantity on axes, with negatively inclined curve.] Total utility proportional to areas bounded by ordinates established at point of the quantity in question. [Discussion of change as result of adding dx.]

Historically, Austrians confused rate and increment, using <u>Grenznutzen</u> for both, tho generally for rate. The two sometimes confused in Jevons, but he almost never spoke of increment, always of rate. Wicksteed first distinguished, calling increment the marginal effect or utility and the rate the marginal degree of utility or marginal effectiveness. Marshall had confused, but distinguished in 4th ed., following Wicksteed. Must distinguish, for at equilibrium position last change in acquired must equal last change in lost [In brackets: "Due to choice of units, says I."], but marginal degrees of utility not equal, but proportional to prices.

Difficulty of finding rate when not uniform for all increments. Another, relating to units of anticipated satisfaction. Used dollars worth, measured by invariable unit area. Not invariable, but varies with stock of dollars. Lose one and rest have higher importance. Simplifies.

Only way out to set up arbitrary unit representing always same utility. Purely arbitrary; must be assumed; commensurable for different commodities <u>for same person only</u>.

I. Fisher justifies saying possess right to 100 loaves of bread in given period and in doubt as to whether to add another or buy oil, and toss a coin. Then have equality and can take as unit, but how transfer to other goods?

What relationship between subjective value theory and price theory?

[In margin, "Says Jaffé,"] Wicksteed's "marginal utility" quite consistently is used as rate, like Jaffé,'s marginal degree of utility. Are confusing passages. When Walras says proportional, means proportional to rates of procuring utility <u>to given individual, not in general</u>. If give 60 eggs for 5 pounds of tobacco and stop, at exchange rate of one dozen/one pound tobacco, stop because feel the rate of procuring utility through eggs is 1/12 of rate through tobacco. Is the proposition to be proved. If so, is subjective and only to be felt by the individual.

Can't regard price as related to marginal degrees of utility for different individuals.

Does rich man give up as much with $200 as poorer man with $20? Differences of income may make impossible to transfer units. Also differences in temperament and education.

May sometimes <u>have</u> to make this invalid argument from price to marginal utilities for different persons. Was something of this even in example above to show invalidity. Can't reason on social matters without assuming rough comparability. Consumer's surplus on stretched assumptions.

Thinking of individual, price, and du/dx, no true [?] explanation of price configuration, for both price and du/dx intelligible only in larger pattern. Proportionality between du/dx and price not enough. Marginal utility not "cause." Now mean by causation only a manifestation in the individual event of a general principle. Know cause, we say, when have this manifestation of general principle.

Beginning page 34, Wicksteed says fundamentally as dependent as the marginal costs. Can have cases in which either acts as a cause. [Equation with equal ratios of prices, of amounts consumed, and of marginal utilities, for an individual.] Cannot assume rareté, the cause of price, even tho Walras so speaks.

One prices given by world market, will exchange so as to get maximum satisfaction. Individual can't alter prices, but can vary quantities. Is individual, given set of prices, who can determine quantities exchanged. May think of quantities as stocks or flows. Quantity of any one commodity taken or given by any individual [as] a function of price. Add, plus and minus, and have supply and demand, each a function of price. Given prices, utility functions, and initial stocks and their distribution among individuals: [end?] how much exchanged. If supply = demand at these assumed prices, they are equilibrium prices. If not, drop assumption that are given and treat as variables.

Summarizes procedures of Walras and Wicksell. Knowing quantities, drop assumption of given prices, and no equilibrium until simultaneously have supply = demand, maximum satisfaction, and budget equation (amount given = amount received). Drop givens of quantities, and get theory of production (cost = price). Drop givens of means of production to get capital formation. There the static breaks down and must jump into dynamic.[9]

General equilibrium is of equilibrium between production and consumption by means of exchange.

In above need also technical coefficients and individual's propensities to consume.

Theory of exchange: Have given quantities of goods, not suited to our particular, and given, tastes. Must exchange, and find are ratios to be established. Must start with some random ratio. Two elements in equilibrium eventually: quantities exchanged and rate, or price. [Diagram with unlabeled axes and negatively inclined curve with arbitrary point P, showing area of price times quantity:] Total money spent.

Wicksell says can deduce prices from quantities exchanged only when have two goods and no more than two. Why?

In present universe, may never exchange cheese for autos, and have no ratio. [Alongside in margin, in brackets: "No!" or "No?"] Have to use third exchange.

Two commodities, with given rate of exchange, and how find quantities?

Easiest for two uses for one commodity, and is no difference in principle. Problem same for individual who would exchange apples for nuts as for him who would take apples either raw or as cider. Same problem of dividing stock of applies, but in one case rate of exchange determined by market and another by technology.

Corn (grain) used either directly or for feeding chickens. No added difficulty. Assumed that grain and poultry neither <u>competing</u> nor <u>completing</u> goods (bread and butter). Rate of change of utility of bread an increasing function ["f(g)"] of butter, by Schultz. Competing goods (mustard and ketchup): more of one, the less intense the desire for other. No completely independent goods probably. May conceive of limiting case of independence.

If no increase or decrease of utility in each successive spoon of cereal, and likewise no change in rate of utility as chicken consumption increased, would use <u>all</u> as cereal or <u>all</u> as chicken, depending upon preference for any one spoonful.

But rates of utility may vary with amount consumed. Assume constant exchange ratio of five pounds of grain to one pound of chicken. If has so much grain that couldn't eat all as cereal, some go for chickens, and so until marginal degrees of utility in proportion to exchange values.

Diagrammatic proof: [Two diagrams, incompletely drawn; one with du/dq and pounds grain used as cereal on axes, the other with du/dq and "pounds of grain earmarked for use as chicken = 1/5 of # of chicken."] Only if have these <u>comparable</u> units can do Wicksell's trick of combining or comparing directly the two curves. Superimpose second curve on first, reversing direction of axis of abscissa. Find that pays to give up 28th pound as cereal for fist as corn, etc., by differences in boxes. Gains until #17. No gain, no less there. If further, loss. Will be 11 pounds devoted to chickens (2 1/5 # chicken) and 17 pounds cereal.

Set out to prove marginal degrees of utility proportional to exchange values, and have only equality of marginal utilities.

Former shown: Exchange value between pound grain and pound chicken is 5 to 1, and between 1 of grain and 1/5 pound of chicken, is 1 to 1. When units of quantity the same, numerical value of marginal utilities and marginal degrees of utility are the same. [In brackets: "For certain discontinuities." "Jaffé, agrees."] Marginal degrees of utility are 1:1, as are marginal utilities, as are prices.

Usually mean by price the quantity of money or other good exchanged for <u>customary</u> units. If at margin here, marginal utility and marginal degrees of utility equal, rate of diminution of utility here for one pound grain and one-fifth pound chicken the same. But wish rate for whole pound of chicken. If utility of 1/5 pound of chicken declining at same rate as that of pound of grain, utility of whole # declining at five times the rate of the pound of grain. Not an average rate of five 1/5 # units of chicken, but the rate <u>at that point</u>.

October 18
Footnote in Schultz, "Interrelations of Demand": Calls marginal degree of utility "final utility" and marginal utility "marginal utility."

Two measures numerically equal, tho different in meaning, if dxi = unity.

Marginal utility of Wicksell can only mean rate, from context, save when choice of unit makes equal, and in some passages, as on page 57, where uses ambiguously as marginal utility.

Jaffé, differs with translator and uses his own terms.

[The next six pages of notes are on the back of Smith's papers, "A Prefatory Note on Terminology" (one page) and "The Choice between Two Alternative Uses of a Commodity." Following Jaffé,'s usage the "Prefatory Note" provides the following definitions:

Marginal degree of utility: the rate of change of the total utility of a commodity with a small change of the quantities possessed

Marginal utility: the change in the total utility as a small amount of the commodity is added or subtracted]

What use made of subjective value reasoning in explanation of price configuration? An auxiliary subjective theory necessary?

At equilibrium in exchange, marginal [In bracket, above line: "degrees, Jaffé, would say"] utilities proportional to price.

Simplify by use of two commodities only or by one with just two uses.

First assume independent in consumption and no more trouble in turning grain into chicken than into cereal. Took as given in advance the utility curves for grain and for chicken. [Two very incomplete diagrams, intended to show pounds of grain ear-marked for use as cereal and "as chicken," respectively.]

Reverse direction of axis of abscissa for chicken and superimpose on cereal. [Diagram thus described, with statement alongside indicating "11 # grain = 2.2 # chicken, at technological rate of 5:1." Diagram shows of AB = 28 pounds, AC for chicken = 17 pounds, and CB for cereal = 11 pounds.]

At C, where final increment converted into chicken, marginal degrees and marginal utilities both equal, because choice of units.

Want to prove marginal degrees proportional to prices. Marginal utility of one pound of grain same in either use. Bases of rectangles same; areas same; and heights same. Per pound of chicken, rate of utility must be five times rate of utility per pound of grain, since five times as much grain per unit. (Base of rectangle narrower and height correspondingly greater.)

Man has quantity of sackcloth that can exchange for sick at five yards sackcloth for one yard silk. Distribute resources, assuming divisibility, so satisfactions from last cent spent on each equal, say for 5" sackcloth and 1" silk. Areas of marginal utility same, so altitude of silk five times sackcloth, if base in inches.

May be either <u>completing</u> or <u>competing</u> goods (and may be both at different ranges). Complete if each better with the other than alone. Compete if each worse.

Schultz distinguishes: [Mathematics of independent goods and of goods which complete at one point and compete at another point.]

Competing goods are substitutable. Limit of substitutability when each unit exactly like every other, which is why have diminishing utility.

Schultz says independent if marginal utility of either depends only on quantity of itself. Completing if same quantity of either yields more than marginal utility if used together [sic].

If independent, AB of grain divided: AC as cereal and CB as chicken.

Suppose: if place grain in better fashion get one pound chicken for four pounds grain. Each pound grain = one-quarter pound chicken rather than one-fifth pound. Increment of utility greater from one-fifth pound. Rectangles along AB higher than original ones for chicken, at start, anyway.

If consume same amount of chicken as before, would need to give up only 8.80 pounds of grain. [Diagram, poorly drawn, to show when same and when less than before.] Utility curve itself not affected. Have merely taken the same area and squeezed it up. Have only changed the units of the base.

Total utility of grain increased. May take less or more than before of chicken [In brackets above line: "grain as cereal?"], depending upon elasticity of demand for chicken [In brackets in margin alongside and with arrow: "For grain, if take less chicken?"]. May take more chicken and still use same cereal, in case of unit elasticity. [The mathematics and diagrams thereof, seeking condition of maximum satisfaction = equilibrium.] How far go on depends upon slope of each curve [. . .]

So far, an isolated individual and distribution of commodity between two different uses.

Faced with market, usually takes for granted the price at any given moment just as did the technological ratio. Proportions so will find marginal degrees of utility proportional to prices again.

Define equilibrium: [Mathematics of all marginal utilities proportional to corresponding prices, with one as numéraire.]

Is one of conditions of maximum satisfaction, given prices and utility functions. Unknowns are the n marginal degrees of utility for n commodities. Only n–1 equations, but have also the budget equation. [Mathematics thereof.] Means that marginal degree of utility per cent or per dollar the same in each use (weighted marginal degrees of utility equal).

Individual has corn to exchange for coffee beans. At ten pounds corn per one pound coffee, given stock of corn and utility curves of coffee and corn, will give 100 pounds corn for ten pounds coffee (for six month period). At next period, price to nine pounds corn per one pound coffee. Assume not competing, so expect greater consumption coffee. Corn consumption? Depends upon proportion of increase of quantity of coffee taken and proportion of fall in price.[10] May increase outlay for coffee, having less corn to consume. Demand for coffee elastic, its quantity rising as its price falls. If demand less elastic [less of an increase in quantity]. Supply curve of corn now positively inclined on Marshallian axes. Results from elasticity of demand for other commodity. [Shown diagrammatically.]

These supply and demand curves drawn for stocks at instant of time, not flows from production.

If production increase takes long time, may get for a while a negatively inclined supply curve for short time.

Example: rising agricultural prices letting farmers use more of products themselves.

Falling wage often means greater amount of work offered. Mrs. Robinson says is typical.

Many interesting questions: Can have backward-bending curve? Contradicts indefinite expansibility of wants? Latter is function of time, but at point of time wants are socially, psychologically, etc. determined.

Have assumed known prices and sought equilibrium quantities. Same for one individual and two uses of one commodity, two individuals exchanging with each other, or many individuals in perfectly free competitive market. Point of relative maximum satisfaction where [marginal utilities proportional to prices].

Given utility functions and stocks or flows of commodities, see conditions of pricing. Two individuals in isolated state come to exchange. De Quincey's example of musical snuff-box on Lake Superior and two travellers for the wilds. The final chance to get this luxury. Owner tries to squeeze most possible out of you, and you pay 60 guineas for it, tho easily obtained at 6 in London. Where equilibrium price? Marshall tried in Appendix F, on barter. Concluded rate governed by accident. Edgeworth constructed contract curve to show range of indeterminacy. Wicksell does simply, with plains peasant and sack of corn vs. forest peasant and one-half load of wood. Plain willing 4:1. Forest willing 1 corn : 1 wood. Thus limits marked, and possible to have exchange. Maximum will be given exceeds minimum will be accepted, and within range depends on bargaining astuteness.[11]

[Traces bargaining possibilities within range, saying, "Either way a bargain possible."] Had stocks been greater and only this opportunity to exchange, question of quantities to exchange indeterminate. Prices also within limits. Whatever price is reached, finally, [marginal utilities will be proportional to prices]. Even tho price indeterminate. [Shown and discussed mathematically. Further mathematical discussion of reaching prices proportional to marginal utilities.] See Wicksell's footnote, beginning page 57.

Wicksteed, pp. 474–526.

Cassel, pp. 131–164.

Kaldor. Nicholas, A Classificatory Note on the Determinateness of Equilibrium, <u>Review of Economic Studies</u>, 1: 122–136 (February 1934,), especially pp. 132–136.

R. F. Kahn, "The Elasticity of Substitution and the Relative Share of a Factor," <u>Review of Economic Studies</u>, 1: 72–78 (October 1933).

[A number of pages are missing from the notes. The notes resume with the seventh page for October 25th.]

[October 25]

[Six pages missing.]

[Mathematics of exchange, again.] This the usual solution of problem of distributing resources, given prices. Different solution if even any one price different. Would affect the transposed curve and the value of the total stock in terms of (A). Proves demand for any one commodity a function of prices of every commodity. Demand and supply schedule would need <u>all</u> prices in the price column.

Budget equation extremely important, for tells can't maximize beyond that budget, the limits of your resources.

Whole first of Wicksell deals with value and exchange. Abstracts from production. Cassel does same in equations (1) and (2) of Chapter IV. Holds superfluous to go into psychological basis. [Cassel's mathematical formulations said to look "exactly like those of Walras, but isn't, even if add to get market curves."]

Cassel argues from scarcity – a fixed quantity at the disposal of society. [Cassel's] S_a is quantity in existence, not quantity offered. Da is quantity <u>wanted</u> to be in one possession at given price. When equals amount existing, have equilibrium. Schumpeter says Cassel is 3% Walras and 97% water. Cassel: As price falls the quantity wishes to retain for self increases (or quantity wishes to have). [Mathematics of exchange, again.] Merely redistribution of same quantity among the people. [Equilibrium when] Demand = supply in Walrasian sense.

Cassel establishes [that demand for a commodity is a function of prices of all other commodities]. Established by economic principle, that won't take worse bargain for a better one. Can derive Walrasian from Cassel's. S is given, the initial stock.

Cassel's curve is of the total amount man wants to have at each price.

Walras' is of the amount he would like to <u>add</u> to his stock at each price.

Jaffé, thinks Cassel stupid to throw away utility analysis. Of course, have to take something as empirically given. Jaffé, thinks a more rational basis if go back to utility explanations of choices.

In Marshallian school is a hiatus between utility curve and demand curve.

Walras integrates the utility and the demand analysis, deriving latter from former. [Diagram.] Superimposes price axis on utility graph.

Wicksell passes to second condition of general equilibrium: demand and supply equal.

Drop methodological assumption that prices given.

Want to know if unique, multiple, or no solution for equilibrium of demand and supply. If many, indeterminate. For this is isolated exchange important.

If minimum acceptable doesn't exceed maximum buyer will pay, may be exchange. No reason why exchange should begin at any particular place within those limits. Economic logic tells us nought of course of prices. Skill, economic power and position, etc.

This is practical – wage determination in modern world is an isolated exchange.

Economic condition is to have exchanged such quantities at end that marginal utilities are proportional to prices for <u>each</u>.

[Mathematical example of exchange through bargaining over price.]

Is theory of emergence of equilibrium price, not of <u>determination</u>. Done by groping, or <u>tâtonnements</u>.

Isolated exchange goes, and stops, then goes again on further inducement until one stops, then perhaps again, etc.

[Mathematics of starting initially at equilibrium price and of not doing so.]

Edgeworth's contract curve and isolated exchange: found in Marshall's mathematical note, in Edgeworth's <u>Papers</u>, Vol. II, pp. 126, 131, and elsewhere. [Diagram and discussion of bargaining in that context.] Continue exchange by changing prices. Many intermediate curves. Final points of each form contract curve, the locus of bargains which it is to the interest of neither to disturb.

<u>November 1</u>
Wicksell, pp. 83–100.

Cut class, with permission. Notes from Jaffé,: [?]

<u>November 5</u> [May indicate date received notes from Jaffé,.]

Contract curve: [Diagram and discussion.]

Extreme boundaries [Over line: "?"] are best possible for parties concerned, that get beginning advantage. (Can't be better, if is to be any exchange at all.)

[In brackets, alongside: "?" "Are indifference curves independent of price?"]
   Equation from Marshall's note, first formulation. [Equation.]
   Practical significance of isolated exchange – marginal productivity [sic] theory. This indeterminate theory more realistic.
   Indeterminate because we don't know enough factors. There is a wage.
   Unique, multiple, or indeterminate solutions in open competitive market?
   Moore's maximum of income for each factor (perfect mobility), perfect knowledge (need only know where to get best bargain, says Marshall).
   [Further mathematical discussion of bargaining, with different demand elasticities.]

November 8
John R. Hicks and R. G. D. Allen, A Reconsideration of the Theory of Value, Part I, Economica, n.s. 1: 52–76 (February 1934).
   Paper on marginal rate of substitution in relation to marginal utility. Concentrate on pp. 52–69. Leading up to use of indifference curves.
   Frederic Benham, Economics, Appendix to Book I, "Indifference Curves," pp. 89–100.
   Every demand curve implies supply curve of one or all other commodities (in a 2- or multiple-commodity universe). A neutral money [. . .] may be taken to represent all the rest. General character of demand function [. . .] at each point.
   [Next four pages of notes are on back of pages of Smith's paper, "Derivation of the Demand-Supply Curve," with reference to L,on Walras, The Geometrical Theory of the Determination of Prices, Annals of the American Academy, 3: 50–51, 1892. Jaffé,'s comments include the following: "Well done, but not elegantly done. You might have developed the same conclusion with a neater notation."]
   [Mathematics.] Can derive general law of supply from a demand [. . .] Price of A could never be zero unless supply of A economically speaking infinite (so large that is free good). Directly, at very high price could give up little, but would still sell something in order to satisfy desires.
   Derived this supply curve directly from utility functions, too, as in Walras. However low price of A (high price of B) some of A always demanded (practically enough to satiate if price of A very low), and some of B will be supplied. May be no demand for B at that high price of B, but always an offer.
   Footnote bottom of page 57, Wicksell: Statement is curiously involved, but ideas not so very difficult. Commodities independent in re consumption, and then pq rectangle (for B's supply) must continuously increase, for demand for A continuously increases as price of A falls (price of B rises). [Diagrammatic

elaboration.] When have universe of two commodities, if demands not interrelated (competing or completing), can posit negatively inclined demand curve, whence rectangles inscribed in other supply curve must increase, as are proportional to ordinates of demand curve.

More complicated when drop assumption of independence and let be partly substitutes. Then not necessary, for demand for A may fall as price of A (a positively inclined segment of demand curve) and still greater fall in supply of B possible.

Butter and oleomargarine example [of competing goods]. [Diagram and discussion.] If price of butter rises from six pounds of oleo to seven pounds, being competing, with butter preferred, quantity butter supplied for oleo decreases rapidly, for can get enough margarine for cooking by sacrificing less butter, and perhaps use some butter for finer cooking.

Like Marshall's example of grades of tea, where less taken of lower grades of tea on price fall of all grades of tea.

Dictum that negative slope the one universal law of demand not true for related demands. [Elaborated mathematically.]

How will marginal degree of utility of margarine be affected by change of quantity of butter? [Mathematical analysis.] Total utility increased at a diminishing rate (marginal degree of utility curve is negatively inclined.

Know, at equilibrium, weighted du/dx's equal.

[Where "very small part of income spent on any one of these commodities,] Problem is now in a form that can be statistically tested [to determine if relationship is completing, independent, or competing.] In first case, increase in quantity demanded of B would increase price of A. In latter, increase in quantity demanded of beer would lower price of wine.

Schultz has studied price relationships and deduced interrelationships of demand.

If independent, increase in quantity demanded of B does not affect price of A.

Wicksell supposes (?) manufacturer has large inventory and sees for long time to come no prospect for rise, with perhaps fall in prospect (?).

May sell a part of inventory at whatever price will bring, to pay bills and go on. Of course, too low price may be absurd.

Supply curve may rise steeply [In brackets: "Why not vertically?"] at beginning, become parallel to price axis, and then become falling curve.

If price is too low, may hold part or all of stock (reservation price). Corresponding demand curve for all other things (diagram with demand curve of varying elasticity shown]. In world of two commodities and pure exchange, demand curve and supply curve of same commodity are independent. Not that demand and cost curves independent. Demand for B depends upon availability of A to holders of A and supply of B availability of B to holders of B. If two groups different, complete independence. If interdependent, an infinity of possible positions of equilibrium. [Diagrams showing unique determinations.]

Kaldor deals with equilibria and their conditions: Distinguishes between determinate and indeterminate, unique and multiple, definite and indefinite.[12] Determinate vs. indeterminate: Can think of equilibrium as independent of path of being attained (determinate) or dependent upon path (indeterminate). If arrived at instantaneously, not affected by path. Quite a determinate equilibrium on silver market in London, where bids and offers all sent in and price agreed upon that will clear market, all being known.

If began day by feeling way, or with many [. . .] cases, would find early transactions altering supplies of money, and thus final price of day, which would be affected by the path, by the cost of exchanges during the day.

Marshall wanted to avoid by assuming all transactions affected very small quantity of total money, so $dm/dp = k$.

Walras used <u>bons</u>, provisional certificates. Prices cried out at random, at which are certain offers to buy and to sell. If always at these first random prices $S = D$, will be the equilibrium prices. If not, as is probable, cry another price (closer to equilibrium prices). In this case, final price unaffected by path, for nothing given up during process.

Edgeworth: Re-contract. People buy and sell, recontracting as news changes, etc., until come finally to an equilibrium price, in which case, he says, path does not affect final price. The whole of the transactions must be recontracted.

Aside from these, have indeterminate equilibria. [Diagram.] Get equilibria in cases of indeterminacy, but depend upon starting point and path.[13]

Cases of unique and multiple: Demand and supply curves crossing several times, as in Marshallian downward-falling cost curve. [Illustrated with diagrams. Under some conditions stable; under others, unstable.] Multiple rather than unique equilibria. May be no [equilibrium] price at all.

Have indeterminate equilibrium when demand and supply curve[s] coincide for any part of lengths. Economic data tell us nothing, so say indeterminate [. . .].

In the realm of production, have also the cobweb theory. [Results turn on relative elasticities. Can have convergence and definite (DD quite elastic and SS quite inelastic), or divergence and indefinite with "no equilibrium at all" (DD less elastic than SS).

Important thing to remember is that at any given moment in an actual market, there is a price. Nature determines it, tho we lack elements to explain it.[14]

November 15
Hicks, Economica, pp. 58–69.
   Allen, Mathematical Analysis for Economists, Chapter 5, pp. 107–29.
Demand, total revenue, cost, and other functions (including indifference curves).
   Mordecai Ezekiel, The Cobweb Theorem. Quarterly Journal of Economics,
52: 255–280 (February 1938).
   Main purpose to determine whether or not theory of subjective value indispensable. Answered by showing how Walras and Wicksell derive price functions
from utility functions. Have shown marginal utilities proportional to prices.
Given the utility functions, the amounts possessed, and the prices. Amounts
unchanged, the unknowns could be determined by equilibrium conditions. If
add amounts taken and offered at given random price no certainty would
equalize. Some would fail to maximize satisfaction. No equilibrium. Drop
assumption of given prices and proceed to query whether is a uniquely determined equilibrium price.
   Need supply and demand schedules or curves. Derived from individual's
requirements in order to maximize satisfaction.
   Generally offer starts at price greater than zero, increases, and then decreases.
May add individual offer curves to get total social offer curve. Each individual
offer curve implies a demand for other. If independent, demand negatively
inclined; if not, may be pos[itively inclined] in part.
   Once have demand curves can study equilibrium prices. Is a unique price?
Not in isolated barter, but a whole range, depending on conditions of exchange,
often non-economic conditions. Then to price-determination in barter in competitive market. Analyze relationship between demand and implied supply curve.
   Always from simple to complex, so start with single intersection of supply
and demand curves. Two sorts: (1) Marshallian: P' greater than P,
supply increases and demand decreases. (2) Walrasian: Both demand and supply
decrease at P' greater than P, but demand decreases faster than
supply. Marshallian: A simple case, but not so simple, especially if get into
realm of production. Cobweb theorem there, and perhaps also in pure exchange.
Elementary factors of production are by definition unproduced. Deal only with
stocks. Is any sort of lag between price inducement and response to it?
   Walrasian case: Assume can group all necessities of community into
composite commodity, (N). All others, as one, are called (L). Community must
divide its resources, depending upon relative prices. [Diagrams comparing
Marshall's and Walras' respective characteristic demand curves for N and for
L: For N, going down demand curve from left to right, Walras' demand curve
has elasticity less than one over wide range, unit elasticity over intermediate-
sized range, and elasticity greater than one over small range; and Marshall's is

the reverse: small range of elasticity greater than one over small range, unit elasticity over intermediate-sized range, and elasticity over wide range. For L, Walras has elasticity less than one over small range and elasticity greater than one over wide range; and Marshall's is the reverse: wide range of elasticity greater than one, and narrow range of elasticity less than one.]

[Diagrammatical analysis of the two groups, now called necessities, turning on differential elasticity.] Wicksell says older economists usually left out the latter case, and queer, since demand for commodity that has risen in price may frequently fall in less proportion (as necessity), that is, elasticity less than one.

Against this, all others constitute a group whose relative price has fallen. Their supply in exchange for the other increases (being the pq of demand for necessities) as the price of luxuries falls. Rises with a fall in its price and falls with a rise in its price. Reach an equilibrium that is stable, for as price rises above Po, quantity supplied is greater than quantity demanded. Demand curve (Walrasian axes) cuts from above.

Increase in demand results in lower equilibrium quantity.

If cost of labor the main portion of cost of prod[uction], and supply and demand curves for labor in this position, may be that commodity reduced in output as price (and demand) rises.

These still unique equilibria.

Now multiple: [Diagrammatic analysis, as above, of negatively-inclined supply curve intersecting demand curve at three points, in both the Marshallian and Walrasian cases, one deemed a stable, and the other two an unstable, equilibrium.] Important that are dealing here with pure theory of exchange, where price independent. [Further diagram:] A cost curve here: problem: profit and loss. In Walrasian case, unstable if demand cuts from above; in Marshallian, unstable if demand cuts from below.

Marshall said never used demand and supply curves for market values, stocks of goods, for found if did so, people failed to see that were <u>flows</u> concerned in problems in which he wished to use the curves.

Have multiple determinate solutions.

Still the problem of indeterminate solution in market of many buyers and many sellers. Substitute or rival commodities. The rye-wheat example of Wicksell: A has 800-pound stock of wheat, and wants 1000 pounds of grain for nourishment. The only commodities in existence [. . .] [Analysis of cases.] When commodities considered singly, may easily have substantial independence. In groups, cannot avoid the dependence which different commodities have upon each other. Demands are related, and thus the supplies. [Further analysis.] This a case of indeterminateness. Is also isolated exchange.

Must consider a universe of more than two commodities. If so, one commodity necessarily serves as standard of value, because of arbitrage, even between relatively isolated markets.

Indifference curves: Universe of two commodities and individual having some of one, the other, or both. [Diagram, with quantity of Y possessed on vertical axis and quantity of X possessed on horizontal axis.] On this system of axes, each point represents combination of quantities of X and Y. Reasonable that some combinations preferred to others. More of X desirable, even tho have no more of Y, within limits.

Plane of diagram is base of solid into which are stuck needles of varying height, representing total utility. On plane 0 at 0X axis, the trace of total utility curve for X. Likewise for Y. Total utility as function of quantities of <u>both</u> X and Y shown by surface above this plane. Each dot on plane represents a combination. Where one is preferred to another, use higher needle, and same heights for equal satisfaction. A surface or hill representing utility. Pass horizontal plane through hill, and the intersections with the plane may be projected on the base X-Y plane, as indices (Ii). On any one line each combination is the equivalent of any other. Are differences between the different curves, tho.

On integrability conditions [two very incomplete diagrams]: Vertical section of hill: (may be through ZX plane). Same contour curve projections from different types of hill (and independent of units). Can't find total utility (volume) from contour or indifference curves. Could go in other direction, but not from indifference curves to utility function.

Marginal rate of substitution [with diagram]. Ob's diminishing and Oa's increasing as give up Y for X. [Another diagram:] Give up always same amount of X. Find that quantity of Y needed to replace given quantity of X is increasing. The law of increasing rate of marginal substitution. Slopes of tangents increasing. If hold X constant, and increase Y, Pareto said marginal rates of substitution of Y for X would increase, but may not always be true. Slopes may not get steeper depending on curves. Law of increasing marginal rate of substitution the counterpart of diminishing utility. [Diagram with straight line, curved and L-shaped indifference curves:] in first case, perfect substitutability, constant marginal rate of substitution; in third case, marginal rate of substitution constant at zero: combination of X and Y at P [corner] as good as any other unless can increase both at once.

Elasticity a ratio between relative increase of proportion of X and Y in possession and relative increase in marginal rate of substitution.

[The next six pages of notes are on the back of Smith's paper, "The Marginal Rate of Substitution and Marginal Utility," with respect to J. R. Hicks and R. G. D. Allen, A Reconsideration of the Theory of Value, <u>Economica</u>, 1 (new series) 1: 55–58, 196–198, 203 (1934).]

November 22
Wicksell.

Transition from 2- to multiple-commodity universe. Important thing is phenomenon of arbitrage, indirect exchange. If put down as data:

Country	For Sale	Not Wanted	Wanted
Sweden	Timber	Corn	Fish
Norway	Fish	Timber	Corn
Denmark	Corn	Fish	Timber

Assume three isolated direct exchange markets: (1) Norway-Denmark: fish-corn: $Pf,c = 2$. (2) Denmark-Sweden: corn-timber: $Pt,c=4/3$ (or 1 1/3 units corn for 1 of timber). (3) Norway-Sweden: fish-timber: $Pt,f=1/2$. Denmark could exchange corn for timber in market (2). If so, would get 3 of timber for 4 of corn, as $Pt,c = 4/3$.

But better than indirect exchange. Buy fish with corn in market (1). Get 2 of fish for 4 of corn and, in market (3), get 4 of timber for the 2 of fish. A gain of one unit of timber, by indirect exchange.

Can't last, tho, for will be great demand for fish, by Swedes and by Danes (the latter for exchange). Don't have equilibrium. Equilibrium conditions are:

$$Pt,c = Pf,t/Pc,t = 2/.75 = 8/3$$

When $Pf,c = 2\ 2/3$, indifferent whether make direct or indirect exchange.

Have abstracted from transport costs, etc.

Indirectly, 4 of corn for $1\frac{1}{2}$ of fish and that for 3 of timber, at the equilibrium price.

Illustrates principle of arbitrage, but doesn't depict <u>process</u>. Use medium of exchange, money. Money used in each market. Goes in opposite direction to goods, completing the circuit, if markets not isolated.

If isolated [. . . Norway selling direct to Denmark who then sells to Sweden] could easily pocket difference in cash, getting same corn. Sweden will lower its price, as finds too much fish coming in.

Now equilibrium conditions: To present, important to note that speak always of relative prices – not of fluctuations of money prices, and effects of money.

All transactions immediately paid for. Money not just any commodity, but a counter to assist in exchange. Wants to delay liquidity preference, etc. until later.

Sec[tion] on objections and exceptions to marginal utility theory interesting.

Equilibrium conditions met in actual market situation only if demand and supply functions continuous (always some change in quantity for any change in price). Utility functions, if for any change in quantity possessed, however small, have some small change in marginal utility.

Roughly, marginal utility theory assumes no breaks in these curves. There are breaks, at very high prices when some cease buying altogether, at very low where some enter market suddenly.

[Marginal utility proportional to prices] has no meaning for those who stop buying. No longer have any of commodity. Not insuperable obstacle, as is only a constant, which can be so handled.

Utility curve frequently broken. Often must abandon it, as when commodity is a unique indivisible unit. Simply a case of isolated exchange, and indeterminate within limits.

What of large units, but not unique?

Böhm-Bawerk's horse market: [hypothetical buyer and seller valuations]. All appear simultaneously on market. All horses are of same quality. All buyers and sellers know the state of the market. [Detailed account of hypothetical path.] Böhm-Bawerk concludes that price determined within latitude of last buyer exchanging or final excluded seller (upper limit) and (lower limit) last seller exchanging or first excluded buyer. [Spelled out in terms of hypothetical details.]

If substitute name, marginal pairs, get formula, market price determined by subjective valuations of marginal pairs.

Wicksell says only for these that have rough equality of price and marginal utility.

If not all like horses, can translate qualitative differences into quantitative terms. Horse no longer indivisible, and have semi-continuity. Prices proportional to increments of horse.

Economically, have many increments, tho physically only one horse.

Edgeworth criticizes Böhm-Bawerk: Not always appropriate to so watch a particular couple. How if weakest actual buyer A_1, for a second horse? His price for a second horse greater than A_6 for a first. Can still couple, but not appropriate to divisible commodity (which Böhm-Bawerk didn't say it was). Neglect the more important problem of infinitely divisible commodity.

If data different, might have been different conclusion. Might have had very wide range, if sellers would pay £10 and buyers sell for œ1. [Smith twice questions the content after the comma.]

Weakness that doesn't illustrate that each part on one side (in concert with some on other side) free to vary quantities. (Again not very apt.}

Doesn't exemplify the law of marginal utility, which it was brought forth to illustrate. (This is a fairly good criticism.)

Will begin next time with Walrasian handling of discontinuous curves, relating to continuous.

INDIFFERENCE CURVES

Henry Schultz, Theory and Measurement of Demand, Chapter I, "Demand Curves," pp. 15–31.

Antonio Osorio, Théorie Mathématique ... de l'Echange, Chapter 8, pp. 302–315. [In margin: "Do not review."]

Irving Fisher, Mathematical Investigations in the Theory of Value and Prices. ["Have read pp. 64–89. Do not review."]

R. G. D. Allen, The Nature of Indifference Curves. Review of Economic Studies, I: 110–121 (February 1934). [Do not review.]

Essentially, difference between earlier analysis and this, is that earlier regard utility as function of quantity of this commodity only. Now take it as related to other commodities.

Utility surface explained, assuming universe of two commodities (for geometrical ease).

Earlier, also, attempted to add all utilities, and maximize utilities by maximizing sum. These areas probably not additive. Well if can avoid, and can determine maximum without such addition of undefined units. Know if had a utility surface, could easily get Jevonian curves, but can't reverse. Fisher shows, pp. 68–69. If have solid, with X and Y on vertical planes, have the utility curve as function of quantity, whence can get du/dq. [In margin: "Partials for each commodity."] But in latter, ordinates proportional to slopes of tangents of utility curve on one of vertical planes. [Discussion of manipulating "plane tangent to any point on hill," showing "rate of change of utility from increment of X if have certain quantity of Y," etc.]

Thus can get Jevonian curves from the hill. Can't reverse, tho, and the hill is the important thing.

Position or altitude on hill determined by income, rates of exchange, etc. Could be a lot of hills that would give same Jevonian result.

Osorio gives conditions governing indices of indifference curves: (1) Any two combinations, choice of which is indifferent, must have same index, be on same curve. (2) Preferred combination must have index higher than that of any combination to which first is preferred. (3) If, in passing from combination 1 to combination 2, individual in question aware of greater increase in satisfaction than in going from 2 to 3, numerical difference between indices 1 and 2 must be greater than numerical difference between indices 2 and 3.

Just as subjective as marginal utility, says Mary Wise, with assent from Jaffé. Schultz says of properties "high degree of probability," not saying knows anything about these curves. Properties not definitely established, but probable.
[...]
Total utility increases with increase in X or Y. Doesn't apply to all cases, as garbage, as here given.

[Discussion of moving to northeast on diagram, "as acquire more of either good, holding other constant;" though notes "Not necessarily." Dy/dx < 0, negatively inclined, "for must have decrement in X if have increment in Y, and vice versa. Must compensate for loss of X, etc.; economic meaning of convexity toward origin, i.e. "As X increases by constant quantity, dx, dy decreases numerically. Compensation diminishes. If X decreases by constant quantity, dx, compensation (dy) increases numerically" etc. Also: "With fixed quantity of Y, get flattening tangents, and slopes decreasing numerically. Needn't be this way, tho generally accepted." "Marginal rate of substitution of X for Y is slope of tangent at point." Examines definition of elasticity of rate of substitution.]

Abba P. Lerner, "Notes on Elasticity of Substitution. II. The Diagrammatical Representation. Review of Economic Studies, 1: 68–71 (October 1933). ["Read no more."]

Hicks [and Allen], Economica, pp. 69–76.

November 29
Paper on Walras article, 2nd part, pp. 54–61. ["Do not review."] Osorio, pp. 331–73.

Walras' corrections to the Eng[lish]; P. 55, 1st line of 2nd para., "labor" should read "persons." 2nd line, after brackets, insert "or movable capital." P. 56, line 12, first word "services," not "labor" [...].

Schultz, p. 5 through p. 46.

Discussion of summary of mathematical treatment in Wicksell, then Walras' discontinuities, and then indifference curves.

[Discussion centers on having total utility function; problem of "rate of change with respect to a quantity of A, holding others constant, etc."; marginal degrees of utility, "similar to Walrasian marginal degree, but function of all commodities."] Given utility functions and prices, the only unknowns the quantities. Are n unknowns and n equations.

Any given individual a quantity adjuster.

How many different unknowns when prices made variable? Introduces n–1 new unknowns. May take price of one as unity (the numéraire or standard commodity) and measure all others in this. Suppose take money with no utility in itself, functioning solely as medium of exchange. Would appear to be n unknowns, but only n–1 independent, as can reduce to n–1 through arbitrage or indirect exchange. [Mathematical discussion of that.] With n–1 new unknowns, the relative prices, must find n–1 new equations. [Further mathematical analysis, using arbitrage.]

After this, Wicksell says, have determined only relative prices, not absolute prices. If money neutral (as is not), absolute prices immaterial.

Objections to marginal utility theory: (1) Based on assumption of continuity and (2) based on assumption of perfect competition. In absence of these, will have a determinate solution? See how Walras handles. Not satisfactory, in final analysis.

Three possible cases: continuous utility against discontinuous, vice versa, and two discontinuous. Deals only with first case, for medium of exchange can and should have continuous utility curve. [Diagrams and their analysis.] In final exchange, utility given up less than utility acquired. In first not exchanged, would [have] utility more given up than acquired.

[. . .]

Thus, in exchange of continuous for discontinuous, at maximum satisfaction, ratio of arithmetic mean of intensities of last want satisfied and last not satisfied by commodity bought to intensity of last want satisfied by balance of commodity sold is approximately equal to price.

[. . .]

Approximate indeterminacy of isolated exchange, but the more the market developed, the greater gradation of quality. We may exaggerate amount of discontinuity.

Other types of discontinuity: One that is serious is producers' goods – no utility at all.

Indifference curves and Hicks' article: three conditions for indexes – if indifferent combinations, the same index; if preferred, give it a higher; if greater change in passing from I to Ii than from II to III, then indices similarly.

Probable properties of indifference curves [amidst mathematical discussions]: Slope negative. Convex downward. Slope increases algebraically at decreasing rate. Curve flattens at each end. Some question as to what happens to slope as quantity of X constant, but Y increases. For position of stable equilibrium at given prices, marginal rate of substitution must be decreasing. If decreasing, even though equalled price ratio, sale of larger quantity would add to satisfaction, so unstable. [. . .] could gain by moving away from point of tangency. [. . .] Elasticity of substitution a measure of curvature, varying from zero to very large values. [. . .]

Quantity of good taken depends upon price of itself, of other goods, of income, and (says Tintner) of interest rate. Latter not generally important.

Income elasticity = relative change in demand/relative change in income.

Equilibrium when a balance between desires and obstacles. Best obstacle is fact that can't get something for nothing – price. Price sets person on a rentier, or path across his utility hill or indifference plane. Proceed along path to highest point, that of tangency with the price line [whose position is governed by income for given period]. Price determines path. Get a demand-supply curve: Indicates at same time quantity of X supplied and of Y demanded. Locus of c is curve of supply of X and demand for Y as function of slope of path followed by the individual (the prices). Demand for Y increases as price of X increases (or price of Y decreases). [Foregoing interspersed amid mathematical analyses.]

December 6
    Schultz, pp. 46–58.
    Wicksell, pp. 100–144.
    Osonio, pp. 373–87 ["Do not review."]

Schultz's equilibrium and exchange related by Jaffé, to Hicks and Allen's articles. Schultz follows the pattern of mathematical economists – one individual and one commodity, and two commodities, two individuals, etc. Fixed quantities available. Drop this assumption and get theory of production.

[Further mathematical analysis, e.g. of derivation of demand and supply schedules for one individual and two or more commodities.]

May abandon use of marginal degrees of utility, if wish, adopting marginal rates of substitution. Defined ((Y) for (X)) as quantity of (Y) that would just compensate for loss of marginal unit of (X).

On an indifference map measured by slope of indifference curve through point at which individual is situated.

[Analyzes effects of changes in price, etc. Equilibrium condition now expressed in terms of proportionality of marginal rates of substitution to relative prices.] Why bother? In terms of marginal rates of substitution is conceptually possible, in terms of an operation that could find indifference curves, but no such thing for utility. Schultz sticks to Walrasian and early Paretian concept. Marginal rate of substitution between two goods must equal ratio of their prices.

To throw light on relation between income and demand, see budget equation [...] [attention to unspent portion; "may tie up to Keynes"] Schultz finds convenient to express prices in a 3rd commodity, say paper money, having no direct utility. [...] So long as price lines are parallel (slopes same), prices are constant, tho income changes. [Curves] same as Hicks and Allen's expenditure curve. Given income, indifference map, and prices, determine equilibrium quantities exchanged. [Elaborated mathematically.]

Expenditure curve may move in many directions. Here moves to N. E., increasing quantities of each. May not necessarily increase each, though may more N. W. (backward sloping) or S. E. [Further discussion of shape and slope of indifference curve and their effects, and other matters.] Economic fact that reduce consumption of some goods as income rises proves possibility. [...] Whenever have inferior goods, as income increases, and marginal rate of substitution diminishes when (X) fixed and (Y) increases.

Are endeavoring to find out how to determine the n–1 prices. Have just shown that are n–1 and only n–1 equations of conditions determining these. Reminiscent of Cassel: Equilibrium between demand and supply, where latter is fixed stock and former is function of price, price being fixed so as to make all of supply be equated by the quantity people want to possess. [...]

[The] fundamental equations of mathematical economics [involve] maximum satisfaction, budgetary equations, and supply = demand.

When price is decreased, watch the process as two steps, one due to increase in real income and one to fall in price of good itself.

December 13
Wicksell, pp. 144–171.
Paper on Walras' Geometrical Theory ..., last part, capital formation, pp. 61–64.
Finish pure theory of exchange and begin production, as in Walras.

Had reached three fundamental systems of equations of mathematical economics from Paretian indifference curves. [Review of material at end of preceding lecture: Maximum satisfaction. Fall in price of Y commodity: increase in quantities of both Y and X:]

Schultz, Slutsky and Hicks and Allen divide shift into two steps, taking account of fact that fall in price of Y [...] involves increase in real income. Has more of both X and Y, and is on higher indifference curve, so increase in real income in every sense.

Increase in real income arising from fall in price of Y supposed offset by compensating reduction in money income. Would [return to original indifference curve] to where could buy same combination at new prices as originally [...]. Money income that apparently represents same real income as before. Not actually, for different price line is tangent to different indifferent curve. Now moves to point of tangency on new indifference curve, [...] increasing demand for Y and decreasing demand for X.

These are the "direct' changes in demand, resulting from change in price of Y if real income apparently unchanged. Means by real income the utility derived from combination. Money income that leaves real income actually unchanged induces to buy new combination considered equivalent to original (on same indifference curve). Money income leaving apparently unchanged permits buy identical combination as original. Apparent money income [...] minus original money income [. . .] may be called apparent loss or gain, according as is positive or negative.

If necessary to reduce money income to maintain apparent real income the same, indicates gain in real income. (Free translation.)

In fact, money income not compensated, so real income rises, and Y increases [. . .] and X increases [. . .] (one may be negative. The "indirect" change in demand. [Part the "result of change in price" and part "result of change in price of Y acting through change in real income.] Have indicated analytically two steps in the change, one due to change in real income and one due to change in price.

Prefers Hicks' approach: Tries to see effect of income elasticity of demand and elasticity of substitution on price-elasticity of demand curve. [Diagrammatic analysis.]

Hicks says [. . .] small increase in income must be spent wholly on X and Y.

[. . .]

Gives notion of relationships between income elasticities and expenditure curve.

Take also relationship between elasticity of substitution and demand curve.

Breaks into two parts, similar to way Schultz handled, but not the same two parts. (It was said.)

January 3
Wicksell, pp. 172–233.

General equilibrium deals explicitly with wider range of variables than partial. Also foundation for more operational approach. Is thus more possible to bridge economics and statistics.

Has long been said that general equilibrium too complete, and not as practical as Marshall's. But from it stems econometric work.

Operational concept extended to economics from physics by Schultz. Is increasing discussion of the operational character of the work being done.

Schultz: Demand as function [prices of all goods, time] referred to in saying deduced by operational procedure for determining meaning of a concept. New attitude toward concept not intrinsic qualities of the term, but the operations required to measure or determine it, as "length." The set of operations by which is determined. If a mental concept, the operations are mental, as those by which determine continuity, for continuous mathematical function [. . .].

Demand that set of operations equivalent to a concept is a unique set, else ambiguity.

Bridgman said meaningless question if no operations by which could answer. Now admits some concepts have meaning, tho no operations, as sentimental one. Some of most important concepts of economics are non-operational. Vain to hope for quantitative approach from such.

Schultz's book makes clear that Marshall's approach less practical than general equilibrium.

In last few weeks have turned to Hicks and Allen, and to Schultz.

Demand curve of Hicks and Allen. Elasticity of substitution and income elasticity of demand describe most important characteristics of scale of preferences. [Examined mathematically. Some points:] Tells how rapidly one must replace another if new combination is to remain indifferent. All parallel income lines indicate same relative prices. Higher lines mean greater incomes.

In Marshallian demand curve, assume income fixed and all prices except price considered are fixed.

Can get supply curve from the demand curve.

If use indifference curves, relation of other prices and income becomes an explicit part of solution, not a mere background condition. Here see how it figures in problem. [Analysis of derivation of demand curve.]

Supply curve: Assume all income converted into Y [...] Amount of Y given for X [... becomes] demand-supply curve.

Distance from other commodity's axis to demand-supply curve represents quantity retained.

[Given "the sort of indifference curves assumed," i.e,, "increasing marginal rate of substitution," further mathematical analysis of price paths, demand curve, demand in relation to expenditure curve (relative elasticity), income elasticity of demand, etc.] Price elasticity depends partly on income elasticity and partly on elasticity of substitution, therefore.

Increase in demand of two parts: increase in real income from fall in price of x and new opportunity of substituting X for Y due to fall in price of X.

Increase in real income concept clearer than Schultz's, but same idea. Know indifference curves probably flatten as approach axes. Larger proportion of income spent for X, the flatter the curve, and the small the income elasticity. [...]

Hicks seems to get different result. Relative importance of two components depends on Rx: The larger Rx, the greater the increase in real income from given fall in price of X. Any contradiction between this and above?

Want geometrical proof for top of page 67, Hicks and Allen. [...]

Want proof next time similar to that given by Jaffé, for footnote 2, page 64.

On the apparently over-determined system of last time: [with regard to sense of n + 1 equations, Lange said to argue one is "not a real equation, but only a definition. System is not therefore over-determined."]

Operational significance of today's discussion: Hans Staehle, A Development of the Economic Theory of Price Index Numbers, Review of Economic Studies, June 1935. Works with International Labour Office, on international comparisons of cost of living. [Mathematics of "a weighted index number, by quantities, Laspeyre's index formula."] Index of changed money income required to keep individual in same state of satisfaction under changed price system would be the true index number.

Ratio of different money incomes with different price systems, yielding same satisfactions, a true index of cost of living.

[Further mathematical analysis of utility/marginal rates of substitution, income transformations due to changes in relative prices, condition of maximum satisfaction, search for "the true index of the change in the cost of living resulting from a change in price."]

A. Wald's Detroit paper, "Criteria for a Constant Preference Scale Expressed in Terms of Engel Curves." Engel curve is expenditure curve, showing changes in amount of given commodity taken as result of change in income.

Given price situation causes given combination of X and Y. Suppose income changes, and prices same, but happens over time. Would show simply changes in quantity resulting from change in income? But can assume constant preference scales over this time?

Wald proves required criteria mathematically for multiple commodity universe, in 32 pages.

Wassily Leontieff indicated simple criterion. If points are such that could not fit family of curves (that indifference curves intersect), could exclude as not constant preference scale. [In margin: "Not sufficient, is it?"]

January 10
Wicksell, pp. 233–299.

No exam, but to write a final essay (due two weeks from today). Take last section of Wicksteed's chapter on Production, pp. 196–206, using material as nucleus for ten-page essay to include as much as possible of material of course.

Set two problems last time, based on pages 66 and 67 of Hicks' article. First question poorly put, so no answer; second not yet done by himself or any of class.

"Paretian" demand curve last timee. Paareto began it, based on indifference curves. We obsserved its superiority to Marshallian because it makes explicit conditions which are only implicit or stated on the side with Marshall. Best

Marshall could do was to say had different demand curve if income different. Derivation of demand curve from utility curve assumed constant marginal degree of utility of money. Much too simple.

Walras dropped this. Much better. Paretian more elegant, but hardly anything extra. Only good for two commodities, tho.

Income and prices explicit here. Obvious that demand a function of income. Can express variations of demand with respect to income by the expenditure curve.

Demand curve expresses relationship between changes in price and in quantity taken, for a given income.

Expenditure curve between changes in quantity taken and given up.

Inadequate. If money income constant and we confine ourselves to two commodities, fall in price of one means rise in real income. Usually the quantity of good whose price has fallen increases. May decrease. Other good vice versa.

Decrease of price of one good means higher real income, as money income fixed. Total of goods buyable increases.

Increase of quantity of good due partly to increase in real income and partly to substitution of this commodity for the other, to extent that this is possible [...].

Extent that is due to real income change shown by expenditure curve. Extent that can substitute [...] related to curvature (inversely). [Renewed and extended comparison of Schultz and Hicks as to treatment of separation of changes in demand due to price and to (induced) income change, i.e. income and substitution effects, using several different shapes of indifference curves.] [...]

First section of Walras' Geometrical Theory is pure exchange, and corresponds almost to Cassel. Cassel simply asserts that the principles governing pricing are necessary consequences of general economic principle. Walras proves it (maximum satisfaction). Cassel uses it, but will have nothing of term. "Uniform satisfaction of human wants" is vague counterpart of equality of weighted marginal degrees of utility. Sometimes calls it "uniform restriction of wants." Walras starts with group of utility curves representing gradation of desires for added units. Cassel says some sort of gradation of wants necessary. Says income fixed, corresponding to Walras' fixed stocks. Latter's translation into one commodity is really into quantities of "money." Derives demand schedules. Cassel jumps to them. Finds includes prices of all other commodities. Walras proves this.

Get demand-supply curve for B if let price of B vary from zero to infinity while other prices constant at certain levels. If vary one or more of other prices, curve may be quite different.

Similar things true for all other commodities and for all other individuals.

Get system [of general equilibrium for each and for all commodities.] This is like the set of Cassel, but Walras explicitly derives them. Cassel gives as datum. May debate whether to go behind to some psychological implications.

Cassel's demand not quantities individuals would acquire, as Walras' is, but the desired commodity balance (the total would possess).

If define as Cassel, equilibrium defined by total desires for commodity balances = total stock in existence [. . .].

Wicksteed (like Cassel) prefers to define [demand] as total one would possess. Then [supply] = total quantity in existence. Not supply in sense of offers for Cassel, but total available flows in given period. Walras uses sense of quantity offered. [. . .] Supply is quantity given up. Demand = supply for each, but Walras means additional desires and quantities offered. Cassel sees in price a means of equating consumption to stock; Walras sees price movements means equating offers and would-be takings.

Two points of view not inconsistent. Can't find anything about quantities changing hands from Cassel's ["?"]. He could do no else, or would have had to explain offer curves, and had rejected utility. Walras explains by deriving from maximum satisfaction curves.

January 17

For concluding paper mentioned last week, desires complete theory of existence of production and exchange in ten pages.

To round off course, to give us paper prepared for Oxford-London-Cambridge seminar. Asks corrections and questions. On Walras' theory of capital formation.

Walras much neglected in history of economic theory. Affinities with Keynesian counterparts.

Arthur W. Marget, Léon Walras and the 'Cash-Balance Approach to the Problem of the Value of Money, Journal of Political Economy, 39: 569–600 (October 1931).

Marget, The Monetary Aspects of the Walrasian System, Journal of Political Economy, 43: 145–86 (April 1935).

Discussion of Walras' cash balance and Keynes' book on money.

Oskar Lange, The Rate of Interest and the Optimum Propensity to Consume, Economica, n.s., 5: 12–32, pp. 20–23.

Walras' capital and interest theory overlooked. Keynes refers only once, as far as known, and then to his geometrical appendix, which is not a fair view of the text. In geometry can treat only two or three variables at a time; the essence of Walras is his treatment of many variables.

Very few in England, with English names, have written on him. Edgeworth refused to publish parts in English on excuse that would be unjust to rest of book.

[These notes are on back of Smith's six-page paper, "New Capital Formation."]

Not as great difference between Walras and Marshall as might think. Walras thought Marshall also working along his lines, to demonstrate the unity of the whole system of pricing, distribution, etc.

Capital accumulation's setting in Walras similar to Marshall's approach. Walras difficult. Trying to set into a static theory. Many lacunae found also.

Reaches theory through pure theory of exchange (remarkably similar to Marshall's temporary equilibrium), theory of production (normal short-period equilibrium), and then the theory of capitalization and credit. French word <u>capitalisation</u> might mean either capitalization or capital formation.

To this point a moneyless economy, for <u>numéraire</u> merely an accounting unit. No money or interest to this point.

Monetary phenomena affect total equilibrium already established through cash balances, the demand and offer for which must be equal and are kept equal by interest rate, properly speaking.

Discusses conditions and consequences of economic progress. Here we first encounter marginal prod[uct]. Formerly provisionally assumed fixed technological coefficients (no progress, fixed tastes, population, factors, etc.). Logically incongruous to consider as variable before this. [In brackets above the line: "Not from individual view, tho from social."] Nor could have put marginal prod[uct] in chapters on accumulation, for there described only the mechanism, not the reasons for it (came in chapter on progress).

If state of arts constant no reason for changing coefficients unless population grows and land fixed in quantity. With scarcity of land, reasons for capital accumulation.

Walras defines equilibrium and shows how mathematically <u>determined</u>. Jaffé, passes over theory of emergence <u>ab oro</u>. (Does not start <u>ex nihilo</u>.)

Conditions of equilibrium compose the starting point: Given number of individuals, characterized by marginal degree of utility functions, with <u>rareté</u> function of the single commodity only. [...] Each individual further characterized by the resources in his possession. Unknowns are prices of consumer goods, prices of productive services, quantities of consumer goods demanded, and quantities of productive services offered. Also budget equation: total receipts = total expenditures. Given Say's law, [has] a sort of residual equation. [Discussion of numbers of equations and of unknowns.]

Walras' whole theory of capitalization and capital accumulation is to extend this system to determination of prices and quantities of capital goods.

Jaffé, says o.k. for old capital, but not for new

Why a capital goods market? In static scheme, by assuming some spend more than incomes and others less, thus requiring transfer. Otherwise only market for services of capital goods.

What would pay for capital good? Hypothesis that no direct utility of own, so can't use maximum satisfaction and budgetary.

Capital goods if demanded at all wanted for incomes, specifically, net incomes, consisting of productive services.

[Abbreviated discussion of gross income from capital good, price of capital good, rate of depreciation, net income, rate of insurance.]

Is rate of net yield, not rate of interest. Can't call rate of net yield the rate of interest, for latter determined on money market.

[A]nswer as to value of capital good[:] Uses rate of net income in capitalization here, not rate of interest. Determines rate of interest on a separate money market.

If rate of interest greater than rate of net income, entrepreneur discouraged from borrowing, leaving excess of cash balances, leading to fall of interest rate.

Above have capitalization in familiar sense, but explains nothing new. Says not why a demand or supply for capital goods.

If no savers in any sense or spendthrifts, everyone living within and up to gross income, and (2) no capital goods replaced when lost or worn out. Gradual reduction of capital goods. (1) No capital goods market. [Reverse sequence as in original.]

Demand for capital goods a necessary condition of static state, for dissavers are those who fail to provide for replacements. Do not make replacements for love of the utility-less goods, but for fear of the loss of incomes.

To keep income intact in future must distribute present gross income ... among ... production services retained for own use, ... consumers' goods sought, and ... productive services for replacement. [Mathematical analysis centers on "e" or "excedent," gross receipts minus expenditures, said "not savings properly speaking," "A condition of statics for individual," and that "Walras arbitrarily assumes in progressive state that left member (e).] This the condition of economic progress, for society.

Neither provision for maintenance or for new capital as arbitrary as seems. Can subsume both under demand for additional net income. Point at which is zero defines static state.

Nothing such in first three editions of Eléments. Used an "empirical" function for savings. Said to derive would have to look at utility in new light, and cf. present and future.

In 4th edition deals with <u>net</u> savings with an equation among his others. Drops all reference to difference between present and future, but does not mean repudiated idea of time. General equilibrium theory is a cross-section of a moving complex of production, exchange, capital accumulation, and circulating media. Would have been irrelevant to include function of lapse of time. Wrote to Böhm-Bawerk that could not take difference between present and future as a datum, for that is a variable, a function of all other things. Did not preclude Walras from taking preference for future income as datum. Introduces savings function by inventing imaginary commodity (E), of perpetual annuity shares, entitling to one unit of the standard commodity per year forever.

[Discusses capitalization, concluding,] Have determined nothing yet, but are defining a set of relations [in which interest rate is reciprocal of price of "e", said to be] analogous to the previous. Do not yet know what price of e is. Same as formula for discounted present value of permanent series of incomes, so time is implicit. Can express total income in terms of units of <u>numéraire</u>, and thus number of shares as their income.

Number of shares not dependent on price of e, as defined a share as the source of one unit of income.

An income from capital goods equal to the number of shares (qe) giving that income.

Have given net income, and also <u>a desire</u> for net income, felt now, tho income to be received in future.

The more net income, the less intense the desire for an added unit. A marginal degree of utility function for perpetual annuity shares. With regard to income and general price constellation, individual decides to acquire a certain amount more of these shares and may decide to offer some, until equality of equations of maximum satisfaction extended to these.

Get next <u>decisions</u> to purchase, but not actual purchase, for those would disturb the whole thing. If demands some of shares, quantity demanded times price (depe) must enter budgetary equation.

Equations show no antithesis between A. Smith's desire to better condition and Walras' maximum satisfaction. An answer to Frank Knight. At any moment of time can only undertake to better condition within limits of resources at disposal.

Amend other equations as underscored to include this. Like all demand functions, becomes an offer function at sufficiently high level of price of e. [Marshallian analysis of market for e.] Offer segments of these curves are really continuous functions. [In brackets above line: "continuation of demand function?"] Hence intersection of aggregate social demand and social supply will only give price of share at which dissaving is exactly equal to saving.

Are now beginning to determine something, [necessary rate of net income] in static state.

To get demand for new shares subtract supply from demand geometrically. This the curve for net demand for <u>more</u> perpetual annuity shares.

Can translate demand curve for new perpetual annuity shares into supply curve of savings, and a function [centering on rate of net income: $i = 1/pe$]. Decision to purchase [quantity of] units tantamount to decision to save [at certain level of rate of net income]. And so on.

Supply curve for savings.

To find equilibrium rate of net income that will correspond to equilibrium rate of capital accumulation, must have demand for net income. Walras gives, from manufacture [mfr.] of capital goods. [. . .] capitalize value of new capital goods = cost of production.

Appears o.k. Added as many equations as unknowns.

Only one trouble. Not clear where demand for capital goods comes from. Seems adventitious, neither directly nor indirectly related to the utility functions, the primary motive of the system. No clue to decisions to manufacture or invest in new capital goods. Demand for more <u>savings</u> than needed to balance dissavings not rationally explained by Walras, tho for that portion is rationally based. If a correct view, the capital accumulation theory still indeterminate.

* * * * *

GENERAL EQUILIBRIUM

Assume:

(1) tastes constant
(2) techniques constant
(3) factors of production homogeneous
(4) perfect competition, a perfect market, pure competition
(5) perfect mobility of factors of production

Universe of (1) individuals and (2) firms.

Likely to be misapprehensions: Order of presentation has nothing to do with totality of phenomenon.

(1) Individual in a universe where is no production, but a certain amount of goods exist – pure theory of exchange. Stock available is fixed, and demand exists through price mechanism.

$$Da = Fa (pa): \text{ partial equilibrium.}$$

$$Da = Fa \ (pa, \ pb, \ \ldots \ pn): \text{ general equilibrium.}$$

$$Da = Sa: \text{ equilibrium condition.}$$

Da is quantity one would like to have in his possession, not only the quantity exchanged.

$$Db = Sb$$

$$Dn = Sn$$

Desire to hold a certain quantity is a psychological fact. Assume each tries to maximize his satisfactions within limits of resources. Not satisfied until feels that can't improve position by making more exchanges.

Final condition: prices proportional to marginal degrees of utility. Not worth while to sell one cent's worth of one good to buy one cent's worth of another good.

Value (in terms of one of the goods) of the total quantity of goods held is a constant. Budget condition.

Above is pure theory of exchange and economy of individuals.

(2) Drop assumption that stocks of goods are fixed:

Let S's be rates of flow in time, variable within limits determined by the plant in existence. Plant can't be increased or decreased. No increase or decrease in number of individuals is possible. Stocks of factors of production for society are fixed. [In margin: "Community as a whole."]

Equilibrium now depends upon additional conditions.

Firm seeks equilibrium in terms of maximum profit. Problems of (1) size of plant and (2) combination of factors.

Optimum size where MR = MC. Perfect competition means

$$MpR = ATC$$

For any given optimum size of plant can have a variety of properties of the factors of production.

Here we come to the marginal prod[uct] theory, for only here does it have sense. Any given rate of output is obtained at a minimum cost for the individual firm when the prices of the services of the factors of production are proportional to the marginal productivities of those factors. [In margin: "Dependent upon sort of production function?" – to which is added: "No. 1-18-44"]

Marginal prod[uct] is rate at which the total prod[uct] increases with an increase in the factor of prod[uction]. A physical concept.

Now Dservice = Sservice is equilibrium condition.

Change in price of service affects marginal cost rate of operation, and proportions in which used.

No equilibrium until full employment under this picture, so far.

This the theory of production. Comprehends the larger part of what has commonly been called the theory of distribution.

(3) Have still to consider the value of the factors of production themselves, as distinct from the value of the services of the factors.

Rate of capitalization indicates relationship between price of service and price of factor.

Rate of capitalization = 1/i. Simple if can find i (and equalize risk).

Some factors reproducible. Drop assumption that are fixed in quantity. As these become a flow or a rate move toward dynamic economics.

When capitalized values of services of factor are equal to costs of production of factors, equilibrium.

Variables: costs of production of factors, rate of capitalization, and previous variables.

Theory of capitalization above.

(4) Theory of money. Depends upon whether have metallic or paper, etc.

If metallic, price of metal in coin = price of metal in bullion in free market.

A picture of equilibrium that at any point of time represents a goal toward which we move.

But, market itself doesn't know the equilibrium point and are all sorts of disequilibrium transactions, which themselves change the goal toward which are moving.

Attempts to reach equilibrium change the equilibrium.

(1) Exchange (Pure)

Individual maximum satisfactions within limits of resources. Prices move so quantity wish to hold = quantity in existence.

(2) Pricing of services of factors of production, based upon search of firms for maximum profit or minimum loss. Prices of services so quantity used = supply available.

(3) Prices of factors determined. Equilibrium when rate of capitalization such that price of each capital good = cost of production of each capital good.

(4) Money and credit. If metallic, value of metal in coin = value of same amount of metal as bullion.

NOTES

1. In later years, some economists were prone to speak of the "distorting" effects of particular government taxes and spending programs. Such discussions usually involve general-equilibrium considerations, but also more: (1) Such discussions take as given all other actions of government, including taxing, spending, rights determining, etc. (2) Such discussions presume the optimality of resource allocation prior to the spending or taxing act in question, which is to give effect to both some notion of general transcendent economy and the illegitimacy of the change in governmental activity constituted by the spending or taxing act in question. (3) Properly constituted, the analysis should permit comparison of the benefits and [e.g. opportunity) costs of alternative scenarios – though such comparisons are always profoundly influenced by ideology – and interest-driven identifications and weights. Strictly speaking, general-equilibrium analysis only identifies consequences of changes/actions/decisions, ceteris paribus; their evaluation is a function of antecedent normative premises. The same applies to consideration of socio-economic and institutional elements.

2. The two preceding paragraphs imply that insofar as a general-equilibrium model/analysis omits certain variables it is not "truly" general, i.e. incompleteness and unrealism is inevitable.

3. Interestingly, included on the list are works – such as those by Schumpeter, Cassel and Moore, though not by Pareto – that contemplated a wider range of variables and a more broadly defined central problem than eventually became true of the mainstream of neoclassical economics.

4. By the 1990s, it became generally agreed that general equilibrium theory, however useful an analytical tool (and that was a matter of controversy), did not apply to actual economies. It is possible in these notes to read an essentially instrumentalist approach by Jaffé, rather than a definition of reality, but the notes are fundamentally equivocal whether Jaffé, intends to be understood as providing a definition of reality or deploying conceptual tools.

5. The institutionalists lamented both the limited number of variables and the choice of central problem. Not all institutionalists had the same attitude toward mainstream economics and toward the relation of institutional and neoclassical economics. Some were undoubtedly bewildered at what they perceived as the neoclassicists' narrowness, some were undoubtedly angry at the failure of neoclassical economists to either recognize and/or acknowledge publicly the perceived normative/ideological bases and/or uses of their analyses, and some who agreed or sympathized with the limitations/narrowness imposed on methodological grounds were dismayed and critical that neoclassical theory was never taken beyond its perceived narrow first stages. Jaffé, had written his doctoral dissertation at the Sorbonne on Thorstein Veblen and undoubtedly had had his fill of institutionalism.

6. This implies a pragmatic theory of meaning in language.

7. These are, of course, a matter of subjective judgment and/or normal (in the Kuhnian sense) disciplinary practice, subject to criticism, but no less necessary on that account.

8. Jaffé is recorded here treating utility theory as a tool, in contrast to its status as a fundamental explanatory principle, i.e. as part of the definition of reality, by the Austrian school. He also distinguishes statements of equilibrium conditions from causal interpretations thereof.

9. One can interpret the set of practices described here and below in several ways: (1) the set of partial models which a general-equilibrium analysis could encompass; (2) a disciplinary practice of puzzle-making and -solving to the exclusion of analysis of the factors and forces operative in the actual economy – though with some sense that actual situations were being studied; (3) a disciplinary practice of puzzle-making and -solving constituting, willy nilly, a prelude to the study of the actual economy, i.e. the inevitable and necessary working out of theory (rationalism) to complement the working out of empiricism, with each informing the other – notwithstanding the belief of many economists in the superiority of pure theory.

10. The notes thus report Jaffé's lectures using completing and competing goods instead of Hicks's complementary and substitute goods, and without explicit benefit of Hicks's analysis of income and substitution effects. But see below for use of substitute goods and for equivalent of income and substitution effects.

11. The indeterminacy aspect of Edgeworth's contract curve has, at least in recent decades, been stressed; indeed, a conventional neoclassical criticism of institutionalist bargaining-power theories of wages was/has been their indeterminacy.

12. The reader is referred to the materials, including Smith's notes on Jaffé's course on Marshall, in Volume 17 of this publication, having to do with the working out of the neoclassical research protocol of unique determinate optimal equilibrium solutions – driven in part by a conception of science which stipulates unique determinate solutions. Notice the ensuing discussions of ways to avoid, by assumption, indeterminacy and non-uniqueness of equilibrium. The problems of indeterminacy and non-uniqueness resemble an itch which neoclassical economists have been unable not to scratch; or, to change the metaphor, a thorn that they have sought desperately to remove, by one means or another.

13. There is no reason why models cannot be constructed which would include starting point and path, and not consider them external complications. Such would yield an array of results, e.g. a family of demand curves.

14. Although it is not clear what is meant by "nature," this is likely an example of "nature" used in the face of ignorance, i.e. as a rhetorical "explanatory" device.

VICTOR E. SMITH'S NOTES ON WILLIAM JAFFÉ'S SEMINARS ON KEYNES, SPRING 1939

Edited with Commentary by Warren J. Samuels

The following notes were taken by Victor E. Smith in a seminar course, Economics E1, on John Maynard Keynes' *General Theory* given by William Jaffé at Northwestern University during the Spring semester of 1939.

Included is a session of the seminar in which presentations were made first by Abba Lerner and second by Oskar Lange and, in other sessions, by Michael Heilperin and Arthur Schweitzer. (Schweitzer's first name does not appear in the notes but he published in 1941 *Spiethoff's Theory of the Business Cycle*, Laramie, WY: University of Wyoming Publications, Vol. 8, pp. 1–30.) Not all of the materials covered in the seminar pertained, either directly or indirectly, to Keynes.

The same approach to editing is followed here that was employed in preparing Smith's notes on Jaffé's lectures on Marshallian economics published in volume 17 of this annual (the present notes are not numbered by date and page, as those were) – for example, editorial comments are placed in square brackets. Unlike the Marshall notes, these include many recorded questions and comments from students; the class was, of course, a seminar. While statements identified as Jaffé's arise in the context of discussion, one can presume that statements not otherwise identified likely record Jaffé's presentation. However, for reasons examined in the introduction to the notes on Marshallian economics, one has to be diffident in attributing statements recorded by Smith to either Jaffé or others or Smith himself (none are attributed by Smith to himself; one wonders

Documents on Modern History of Economic Thought, Volume 21-C, pages 57–109.
© 2003 Published by Elsevier Science Ltd.
ISBN: 0-7623-0998-9

if he was too busy taking notes to participate). The notes are a record of what Smith recorded as having transpired and having been important enough to record, and are useful in that regards. My understanding – which may be imperfect and may not apply to all notes – is that Smith's approach to note-taking was to make a record of what transpired, not merely to take note of what particularly interested him. The dividing line is not always clear, inasmuch as he did not take stenographic notes; still, he undoubtedly did not record everything.

The notes – with due regard to the fact that some of the voices reported were those of students – provide insight into the early understanding – and difficulty of understanding – and interpretation of Keynes's theories (not everyone thought they were revolutionary, but this is in part a matter of how they understood what he was saying in relation to earlier ideas) and how they were compared with, contrasted to, and even combined with other economic ideas – even how "classsical theory" was restated, and perhaps reinterpreted, in more or less "Keynesian" terms, at least in response to Keynes. (The seminar was held, of course, before the highly influential books on the meaning of Keynes, by Lawrence Klein and Dudley Dillard, were published.)

Materials such as these help enable historians of economic thought to penetrate the past and escape the tendency to interpret the past in terms of present interpretation and thereby to identify the *history of interpretation* of ideas: There is what Keynes, or Marshall, said; there is the history of what they were understood to be saying; and there is the history of what their impact was and meant. And there are the varying stories told of what the classical theory was and of its relation to Keynes's theory – given one or another story of what Keynes's story was. In all these respects, the reader will be well advised to hold in abeyance his or her personal conception as to the correct position on all these issues. The interpretive situation in 1939 (no more, perhaps, than sixty years later) was very fluid and should be understood to have been so.

The notes also indicate something of what George Shackle emphasized as the hold on people's minds of received ideas – and their insinuation into the understanding of new ideas. As with Smith's notes on Marshallian economics, one finds latent here the question whether any particular discussion is a matter of an instrumentalist conceptual model or of the actual economy or of how to move from one to the other. The question arises, for example, in regard to the discussions on theories of interest and of capital, where the issue is not necessarily which theory is correct, and the others wrong, but in what respect each theory may, by abstracting from other aspects of the larger subject, be analytically useful – which is one way of making John R. Hicks's point that no one theory can answer all the questions we might have.

Finally, the notes seem to me to suggest – perhaps to confirm what we already know from other sources – something of the state of confusion in economics in the late 1930s, a state which was partly, significantly, but still only partly, remedied by Paul Samuelson's "neoclassical synthesis" and a state which – I am tempted to say "if truth be told," but it is a matter of judgment, not truth – returned with a vengeance in the later 1980s and 1990s. A major caveat to reaching the conclusion of a state of confusion is that the notes record a seminar: (1) presumably the students had had prior training which was taken for granted; and (2) seminars are more free-flowing and explorative than lectures in regular courses.

I am, once again, indebted to Margaret Smith for her help in correcting my transcriptions of her husband's notes.

KEYNES

Seminars by William Jaffé
Notes by Victor E. Smith
Edited with Commentary by Warren J. Samuels

Start from classical theory of determination of interest. Schedule of savings as function of the interest rate and demand for savings a function of net product of investment. Usually thought of as monotonic. [Two diagrams: One with positively inclined curve relating savings to interest rate, and the other with a negatively inclined curve relating demand to marginal product of capital per dollar of capital.] Superimpose curves and have solution, as any price problem. If assume correct, policy to increase employment should shift both curves to right. Does not mean necessarily increase of capital, but only more money invested. Price of capital goods may not remain same.

What would move to right? Laws adding security, so save more at same rate. Guarantees against inflation, increased taxation, etc., for supply curve. Assurance of higher marginal product.

Keynes rejects whole logical structure, as meaningless. Demand for <u>funds</u>: [Diagram with demand and supply curves and interest rate on vertical axis.] If interest rate is higher than equilibrium rate, amount demanded is less than amount supplied. Inconceivable, for amount actually saved and actually invested must always be equal.[1] Total income of society (Y) equals C + I, the incomes from consumers' sales and investment output. Then I = Y – C. Savings defined as excess of total income over consumption, S = Y – C, so S = I.

(Craine remarks that may be different periods.)

Irrelevant that individual may save without investment. If doesn't spend for consumption nor invest, reduces total income of community by just that amount, which accounts for the saving without investment.

What one man hoards another is prevented from hoarding.

I is quantity of money spent on investment goods for given period of time. (Similarly for consumption goods.) Grant this, and can't longer argue that an increase of savings in schedule sense creates immediate disequilibrium. Savings equal investment whatever their amount. Likewise, if shift of demand curve. Keynes rejects idea that supply and demand curves independent.

(Secrist: Question is as to the identity of beginning, $Y = C + I$, which we had by definition and thus by fiat.)

Not told of effect of new investments upon total income in next period. This effect developed by theory of multiplier.

Oskar Lange, in February 1938 Economica: The two theories really only two special cases of a more general theory, which he develops. Choose between assumptions on basis of relation to reality.

Deibler: The reasoning uses a money concept. Believes the entrepreneur thinks of capital goods when borrows.

Mary Wise: Increased investment over amount of saving. Must admit that $dC/dY < 1$ and positive, marginal propensity to consume.

Multiplier, per Joan Robinson: Discrete periods. If marginal propensity to consume is zero, $S = I$.

If 9/10 investment spent, in first period $1,000 goes into income stream. 2nd day, $900 into stream; 3rd day, $810 into stream, etc. Whole amount eventually saved. (No − limit.)

Multiplier in this case is 10. Effects upon income depend upon marginal propensity to consume.

Multiplier depends upon marginal propensity to consume, $k = 10$ in above case.

$$dC/dY = 1 - k$$
$$dY = dI + dC$$
$$1 = dI/dY + dC/dY$$
$$dy = kdI$$

Say over period of time that S gradually comes to equal I.

Identity formerly mentioned holds for each period, but rather involuntary.

Important thing is amount people choose to hold, which determines eventual increase in income.

Better think of multiplier as a rate. Effect of a particular investment depends upon other factors, notably its effect upon other investments.[2] Business confidence, methods of financing, or marginal efficiency of capital.

Multiplier may change during time, declining as income rises.

General implications of argument that useless investment better than none at all, if have great unemployment.

Raises level of income, benefiting society as a whole.

Jaffé: Take government loan for armaments and show effects.

Wise: Assume employment and investment multipliers same. $2 million, with multiplier of 4; increases income by $8 million, a net increase of $6 million. Ignores effects upon private entrepreneurs. If have full employment in beginning, multiplier very low, and get only rise in prices.

Uses labor-units as measures, and wage-units. At full employment, the whole increase is in wage-units, not in labor-units, and get rise in prices.

Deibler: Must analyze the fundamental assumptions and exact meaning of terms.

Jaffé: Prefers to let terms impress selves upon mind as develops argument.

Deibler: Logic is easy, but would not agree to anything unless knew meanings of terms and agreed with premises.

Secrist: Would like to start with simple definitions and clarify meaning.

Crane: Have to know all of Keynes before understand any of it.

Jaffé: As preliminary issue, is saving a function of the interest rate or of income? Hope to define, Y, C, I, S.

November 16

Jaffé: On last week's questions. Mutual independence of supply and demand curves in traditional explanation. [Diagram with rate of interest or marginal product of capital per dollar's worth of already[3] invested capital on vertical axis and dollars of savings on horizontal axis, with demand and supply curves.] Keynesian theory regards as dependent upon each other.

Ordinary supply and demand curves of single commodities may be considered independent, though, strictly, are not.

Demand with elasticity less than one, and price rises, so more spent on it. Less purchasing power for other commodities, and their demand falls to greater or less extent. Less of them bought, and thus less manufactured, so factors of production used there reduced, and prices of their services fall. This will also affect the costs of first commodity and thus its supply curve.

May neglect these indirect effects if commodity takes a very small part of total income. Effects are, practically, imperceptible.

Not at liberty to neglect repercussions in case of savings curve. Increase of savings in an immediate sense reduces the marginal product of existing invested capital, and thus the demand curve must shift when the supply curve does.

Secrist says cannot argue to total view from the disparate, partial view of earlier case.

Another question on time-period of equation, (1) $Y = C + I$ (by definition), (2) $I = Y - C$, (3) $S = Y - C$ (by definition), therefore $S = I$. (Aggregates for whole economy.)

Choose a period for which these equations are true. None of these are stocks; all are flows.

Deibler asks, What logical difference? Jaffé answers, None. Crane says results might not be useful due to "queer" quantities obtained.

$S = I$ is only an analytical proposition, not a statement about a real world.[4] Not from observation; can tell us nothing new; cannot be wrong. [See note 1.]

Usefulness of mathematical propositions in inverse proportion to their obviousness.

Assume: $Y = \$100$ billion, $C = \$80$ billion for year 1938.

$$Y = C + I$$

	$Y = C + I$
Invest $20 billion:	$100 = 80 + 20$
Hoard $20 billion:	$80 = 80 + 0$

If in terms of different stocks and periods, could not have $80.

Custis: If $S = I$, how can thriftiness do harm?

Deibler: What of accumulations of consumers' goods?

Jaffé: Services of availability, in addition to consumption satisfaction.

Deibler: Is this investment of same order as factory construction?

Jaffé: Yes. In same category as working capital. Are tying up capital.

Deibler: A value concept here, not a goods starting-point. Keynes talks of ag[ricultural] carry-overs as investments. [Second sentence possibly a new paragraph.]

Crane: Wages.

Lerner on Cassel: Too bad that doesn't know the literature, he says.

Keynes says doesn't expect thoroughly trained classical economists to understand. Thinks of his theory as revolutionary.

Jaffé asks whether is really this antithesis.

Wages and unemployment:[5] Classical. When unemployment occurs will be reduction in money wages, higher profits, and expansion of operations, until no more unemployment. Unemployment a temporary, frictional phenomena [sic]. Permanent only if not free competition. Answer: Voluntary unemployment if refuses to take real wage cut. If want to take a real wage cut and have not the opportunity, are involuntarily unemployed.[6]

Laborers not free to take lower real wages. If money wages fall one-half (only costs being wages), marginal costs fall, and prices. Equilibrium when prices have fallen as much as costs. No more unemployment.

Custis: Isn't conclusion involved in premise that only cost is a wage-cost.

If cut and prices fall in same proportion, employment, output, and real wages same as before.

Custis asks what if start with disequilibrium rather than this equilibrium?

Keynes argues that can have an equilibrium in which there is unemployment, and the equilibrium not altered by money wage cut.

What evidence that will be [proportional] cut in marginal costs, if wages cut? Demand conditions. Temporary increase in employment affecting demand in such a way as to cause losses, and return to unemployment.

Money incomes rise, but expenditures less rapidly, so entrepreneurs pay out more than people spend, and suffer losses.

Jaffé: Is a certain quantity of money that people like to hold. Services of approvisionment.

(Break here?)

(Not investment for employment purposes, adds Jaffé later.)

Wages cut, and this factor cheaper than others, replacing them. Others fixed in supply in short run. Reduced earnings for them, which will not be in same percentage as wages. [Added above line: "(Why?)"] Prices of consumers' goods will not fall as much. Real wages fall. Real rewards of other factors rise, temporarily. Their prices will eventually fall as much as wages, and prices equally, so real wages back to original point (and employment).

Others fall, despite increase of total real incomes, which is equal to increase of total real costs. [Above the margin, raising the question of whether Smith copied someone else's notes or is questioning his own note-taking: "(Correctly copied?)"] People don't spend all of money incomes, losses, and again unemployment.

Deibler: What of unit costs? They don't necessarily rise.

Crane: If they fall, Keynes may still be consistent if people "save" enough.

Cady: Does include idle capital and land too?

Bernhard: Criticized by Ohlin for assuming always increasing costs and decreasing productivity.

Crane: Two points for study: Will necessarily be an increase in unit costs when absorb labor after money wage cut?

Hohman: Should we talk about unit costs or aggregate costs? Keynes talks always of aggregate costs.

Wise: Marginal efficiency of investment.

Investors and savers not always same people. The decision to save not automatically a decision to invest. Is automatically a decision not to consume. Then reduces incomes. The fall in income necessary to make I = S depends upon the tenacity with which people hold to their decisions to save. One person's decision to save may force another not to save.

What forces determine investment?

Orthodox theory says marginal prod[uct] of capital and rate of interest.

Income depends upon investment in excess of consumption.

[Diagram with marginal product of capital per dollar of investment (as percentage rate) on vertical axis and investment on horizontal axis, and with positively inclined curve rising from point i on vertical axis somewhat above origin.] Definition relates returns at margin to a stock at instant of time. Based on static assumptions. Can be no net investment, else not static.

Jaffé: Not the return on a flow of new capital. [See note 3.]

Can't be used as showing the development of society, though can contrast two points.

Keynes' marginal efficiency of capital: discounted expected net yield on capital good at margin of use as ratio of supply price. Informs us that is new investment. [See note 3.] Marginal efficiency depends on rate of new net investment as well as upon stock of capital. Rate of net new investment affects supply price. If is positive, raises supply price, and reduces marginal efficiency below marginal productivity.

Deibler: Is this a stock of capital in the value sense?

If increase stock of capital, affects the price per unit.

Wise: Marginal product depends upon value of stock of capital. As price of capital goods rises, on rising demand, marginal efficiency falls.

Deibler: Here a shift from a capital value concept to a capital goods concept. Must be sure that are conscious of the shift when you make it, to validate logic.

If marginal product per physical unit is $6 and rate of net investment is positive, marginal efficiency per physical unit added < $6. [May not be new paragraph.]

Hohman: Are now talking of capital goods.

Wise: Wage-units used, to measure total investment or output by employment. Goods not homogeneous. Labour-unit is an hour's "standard" labor.

Deibler says not homogeneous, either

Jaffé: Could use wheat. Even if labor is more or less variable, aids in further task of considering changing values of money.

Wise: one unit of ordinary labor. Skilled labor assumed more productive in ratio of compensation so weight by (Jaffé's: marginal) wage paid. Crane says must assume perfect competition.

Wise: Total real income will vary as number of labor units applied on given capital equipment.

Crane: Money income can't vary in short time.

Wise: As incomes rise, marginal prod[uct] falls, and might say a different unit, but can term it a change in marginal prod[uct] of capital.

[Diagram with rate of interest on vertical axis and investment on horizontal axis, with three parallel negatively inclined curves labeled C2, C1 and C0, from left to right, respectively, and line representing single level of interest rate, *i*.] As level of capital accumulation rises, marginal prod[uct] falls. For given interest rate, investment less as level of capital accumulation rises.

Variables on which depend marginal efficiency of capital:

$$I = f(i, C)$$

Demand for investment goods a derived demand from consumers' goods.

Keynes says not a simple function, but depends upon long-range expectations, not only present consumption.

Supply of consumers' goods a direct function of C. No cumulative tendency to misjudge state of demand. Simplifies by putting consumers' goods industries in Marshallian short-period equilibrium.

Investment depends on short-period C, and expectations of value of money, attitude toward government policies, expected wars, etc.

Bernhard: Recent monetary writers say was not actual gold ["Au" in original] flows but increased investment that made effective.

Wise: Marginal efficiency of investment therefore subject to very abrupt changes. A "discontinuous" function over time. In depression, marginal efficiency very low (is an anticipation). F[low] may change rapidly, and cause total income of society to fall violently.

Crane: Flow of new investment determine money incomes.

Wise: Keynesian is an equilibrium system, not an attempt to develop activity of variables over time. Given certain anticipations and a certain state of variables, will come to equilibrium in a certain way.

(Have discrete periods, here, if analysis to be correctly worked out.)

November 30
Bernhard on Keynesian Theory of Interest and Historical Development of Doctrine ["Theory"] of Interest.

Interest theory crucial, and Keynes stands or falls on it. Keynes a monetary theorist. Monetary theories unsatisfactory until recently, but better now.

English economists: Earliest tied interest to profits or rent. Stated also that banks could influence until Mill, whose monetary theory very confused.

During restriction period in England, problem of proper limitation of quantity of money. Bankers held adequate if only sound commercial assets required. Economists refuted well, but later were forgotten. Ricardo put no faith in limiting notes by amount of notes.

Bank directors can't distinguish true from fictitious bills. Bullion Report, strangely, often quoted by banking school, since definitely refutes. No limit to demand that sound entrepreneurs would make for loans.

If bank rate lower than rate of profit, no limit to demand for loans, said Henry Thornton, Lord King, Ricardo, and authors of Bullion Report.

Ricardo said banks could lend any amount of money if rate was less than market rate.

Early economists apparently never doubted that banks could regulate rate. Had also the theory that quantity of money could be regulated by interest rate adjustments.

Currency school technically won, but banking school remained in power after the Bank Act of 1844. (?) Bank of England directors wanted no responsibility for controlling rate of interest.

Interest rate theory of quantity neglected until Knut Wicksell revived. Austrians (Hayek, Mises) rediscovered too.

What limits expansion of bank loans? Latter said the exhaustion of supply of wages fund. Regarded as figment of imagination now, pretty largely.

Rediscovery of idea of relation of natural rate of interest and bank rate.

If natural rate equals money rate some correspondence between saving and investment.

Natural rate equivalent to "real," "equilibrium," etc. Swedes have done most by correcting general equilibrium theory for monetary economy.

Myrdal: Theory of prices that includes anticipation is the most analytical econ[omist] can do in dynamic world.

Definitions of natural rate of interest,[7] and modifications:

(1) Return on land – Physiocrats.
(2) Profits on commercial enterprise – Classicists. Didn't analyze structure much.
(3) Rate neutral in respect commodity prices, and equal to rate of supply and demand if no money for holding, all loans in goods. Interest and Prices. Cf. Wicksell,

Lectures: Demand for loan capital equal supply of saving, and more or less corresponds to expected yield on new capital.

Hayek: Demand and supply of capital meet in money form, not natural form, so market [rate] does not equal natural [rate].

Marget: In long run, must be as if real capital lent as is. Wicksell came to recognize greater variability, with money rate reacting on natural rate, and partly determining it.

Robertson: rates of flow of savings and of demand for investment.

Problem is really of causes for divergence of interest rates from natural rate. Two causes: (1) money economy; and (2) dynamic economy. Not mutually exclusive.

(Bernhard speaks of net interest.)

Objections to natural rate: Sraffa – even if postulate non-monetary, difficulties in natural rate concept, if allow any divergence from general equilibrium where all rates equal each other and money rate. Have natural rates for each commodity, and different rate for each different length of time.

When savings in non-monetary finished goods converted from consumption to investment. But not really conversion, for prod[uct] had to be so oriented equally.

Savings abortive if not equal to investment. Investment must be planned ahead to equal the hypothetical flow of savings, even in non-money economy.

Writers: Robertson, Davidson, Wicksell, Cassel, etc.

December 14

Lange and Lerner present.

Lerner: Distinctive contribution of economist is his general view of the whole thing. May not freely extend particular case to general.

Say's Law: Business man sees not enough purchasing power for his goods and can cure by restricting supply. Sees supply too great relative to demand. Say's Law points out essentially that won't work for the whole, as output of one industry is demand for other industries.

Keynes has applied same principle in another way – on problem of wage-cutting as ameliorating measure. [See note 5.] True of any particular section of economy, but untrue of whole economy. Less can be purchased if wages cut all around. Can no longer neglect repercussions.

Cannot defend wage cuts as a remedy by the simple argument above. Keynes' book The General Theory of Employment . . . . Usually assumed employment natural, and not needing explanation.

Analyzes argument that wages fall if is any unemployment, and remedy unemployment. But: If reduced by 10% all around, would reduce costs by 10% at margin, and would pay to expand output. Need to consider demand, but lack time here.

Suppose do not expand production: As result of 10% wage cut, prices might fall 10% and profits also by 10%. Now every business man in equilibrium as was before. Would be equilibrium, if happened.

Suppose do expand output: Real income of society increased (more goods and more employment). Proposition, not disputed, that larger amount would be saved. $S = Y - C$.

How is the income created? Some by making consumption goods (some of business men (profits) and some of employees). [Alongside in margin: "(Limited to wage and profit economy.)"] Some by making investment goods ((1) wages and salaries and (2) profits). Income from investment goods equals investment. Investment good anything not consumed. $Y = C + I$.

Total income composed of C and I. Is more saving as real income increases (people other than business men).

These people increase expenditures by less than business men pay them. [Alongside in margin: "(Shift from real income to money incomes.)"] Therefore business men pay out more than receive, and profits diminish. If previous situation an equilibrium one, losses will accrue and continue as long as business output is larger than before.

Can argue the other way round. If restrict output, real income less and expenditure reduced by less than reduction in income (as save less), so profits increase and expand output to former position.

Keynes argues is a certain amount of investment and employment in these lines. Says little about determinants of investment save that depends on rate of interest. Given certain expectations, invest if rate of interest less than expected return, etc.

$C(Y)$: Consumption depends upon income. Propensity to consume. Assume $C = 9/10\ Y$. $S(Y)$. $S = 1 - C(Y) = 1/10\ Y$. Equilibrium as above when rate of interest equals expectations. Case:

	I	C	Y	S
(1)	10	90	100	10
(2)	20	90	110	20

– presumably but wrong, for won't spend 90 consumption: $C = 9/10\ 110 = 99$.

	20	99	119
	20	107	127

– comes to an end when Y=200

	20	180	200	20

If investment 20 instead of 10, income is 200 rather than 100. Given I and propensity to consume can determine income and employment.

(3) Now let be change in propensity to consume to 8/10 Y:

I	C	Y
10	90	100
10	80	90
10	72	82

End at 50, where 8/10 50 = 40:

10	40	50

This is the main way of organizing the problem by the Keynesian theory. Hence get the practical proposals that increase by: (1) reducing rate of interest to increase investment; or (2) increase expectations (as balancing budget may do, or unbalancing it if people think that good for business); or (3) changing propensity to consume (by redistributive taxation, by government borrowing and payments to spenders).

Any particular action may have reactions in several ways, perhaps in different directions.

The theory provides a framework into which can fit remedies, if know effects on investment and propensity to consume.

Lange: Will try to show similarities to other approaches of theory. Shall confine self chiefly to the monetary theory, the key to the whole system. Most important contribution of Keynes a synthesis of monetary and general equilibrium theory.

Earlier had bipartite theory: (1) what determines relative prices; and (2) what determines general price level.

Synthesis was made by Walras, but few read him that far.

Prices proportional to marginal utilities or marginal productivity, but what the general level?

First attempt in terms of quantity theory of money: $MV = PT$, with T the real volume of transactions.

To obtain general price level, need added multiplier by which to multiply relative prices.

Brought in also an additional unknown, velocity of circulation. V became refuge for ignorant. Could take as datum determined by habits of payment, etc., but know that is variable responding to economic events.

Marshall: As any good, demand and supply. Real problem the demand function. For commodities by marginal utility or prod[uctivity]. Not for money, though, for no direct utility. Marginal utility is marginal utility of what can buy for it, which depends upon price level.

Marshall took want for liquidity as the particular place for money. People want to hold certain proportion of real income in cash.

KR = real value cash balances wanted. (R = income)

KPR = nominal value cash balances wanted.

Demand for cash.

Total demands must equal money in existence.

$$M = KPR \text{ Cambridge equation}$$

$$M = KY \text{ (Y = money income)}$$

$$M/K = PR, \text{ or } MV = PR, \text{ where } 1/K = V.$$

Type of theory of Pigou and Robertson.

If K and R constant, P is proportional to M, as in quantity theory of money. Pigou and Robertson dissatisfied, as statement assumes K and R independent of M, but know cash balances related to incomes, and real income may not be independent of M. Variables not independent. Hawtrey gave up assumption K as constant proportional to M.

What determines proportionality? Keynes comes in with liquidity preference. Is a cost in holding cash balances, the rate of interest foregone. $M = F(i, R)$, with R = real income. Determines demand for cash balances. Really only an application of principle of marginal utility.

This the basis for Keynes' determination of rate of interest. Was known that money rate depends on supply of money (currency school, then Wicksell, Hayek, Mises, Robertson, etc.).

Take R as given for now. [Diagram with interest rates and quantity of money (i, M) on axes, with vertical SS at M, but with notation that it may not be vertical, and downward sloping demand curve.] Demand for cash, if income given. Sum of individual curves = market curve. Intersection DD and SS determines rate of interest.

More complicated if let income vary. Propensity to consume and rate of investment vary with *i* or M.

Change in M, by affecting *i*, also affects real income. Back to Cambridge equation, M = KPR. R no longer independent, nor K, but depend on M.

K explained by liquidity preference theory and R by theory of employment. If know how M affects K and R, know also how affects P.

Assume increase in M. Distinguish cases of: (1) involuntary; and (2) voluntary unemployment; and (3) not named.

(1) if owner of factor of production ready to supply added amounts at existing price. (2) if would work longer or only at higher price. Difference is in elasticity of supply. In (1) is infinite [diagram with price and supply on axes and with flat curve]. In (2) [diagram with two alternative supply curves, one

continuous and the other with a kink. (3) No additional supply at any price [diagram with supply curve commencing with flat segment, with upward turn, and with vertical segment, with elasticity designated, respectively, as infinite, as turning from infinite to less than zero, and as zero.] Effect of M upon Y depends upon elasticity of supply, if effect is to lower interest rate, and increase (usually) inducement to invest.

I rises. If is involuntary unemployment can increase I without raising P. This the only effect.

In state (2), I rises, and demand for factors of production, but higher prices necessary. Partly greater employment and output and partly higher prices. Semi-inflation.

In (3) only prices. Inflation, full or true inflation. This the case of traditional theory where R assumed constant.

Get stage of semi-inflation not necessarily from voluntary unemployment but also from bottlenecks. Are different elasticities of supply, and in some branches may be early restrictions and P rise.

Theory gives more realistic connections between M, demand for cash balances, and income. A sort of second approximation. Shows level of real income is a variable with M.

Traditional theory, assuming M unvarying, really assumed full employment. Really assumed only a reallocation of resources.

Keynes says can employ resources that would otherwise be idle anyway.

Gives synthesis between monetary and general theory.

Change in quantity of money affects interest rate, and therefore also relative prices. Can no longer discuss them first and money later.

Explains also the velocity of circulation, the real volume of transactions divided by the quantity of money. No longer either a datum or a rate of interest.

Causal relationship between saving and investment. Has been assumed that investment determined by propensity to save. Keynes says causal relationship the inverse wherever income can change, as it can.

Saving, consumption, and investment, except for case of full employment, are not competing, but completing.[8]

Practical consequences: Chief difference from Cambridge in view of causal relation between saving and investment. Can both eat the cake and have it, and a bigger one.

Discussion:

K = 1/V. Keynes thinks of labor as only unemployed factor, assuming prices of others fall sufficiently so that they are used. Are difficulties.

Spending $90 million more may reduce business expenditures $100 million, but can't stop where are, so would have government spend enough more to make up for it. Could go on indefinitely.

Borrowing preferable to printing, because people apt to begin spending so rapidly that would require rapid imposition of heavy tax to equalize. People think borrowing better than printing.

Lange likes Hansen's results in his latest book (Full Employment, Stabilization . . ., or some thing of sort) which reaches, practically, the Keynesian conclusion.

Lerner would prefer profitable investments to unprofitable ones, but the question is rather of unprofitable rather than no investment at all. PWA [Public Works Administration] better than dole, because spends more. As long as have unemployment, do not merely take from one to give to others, but give a newly created income or output. Immediate effect of unprofitable investment same as profitable, which is sufficient justification. Future effects differ.[9]

Many government projects less useful than might be, because government feels obligated not to compete in useful things.

Lange agrees with Deibler's argument that lowering of rate of interest may not always be capable of inducing investment, when will be necessary to use public works.

Keynes advocates public works only when are unemployed resources, so that will not be taking them away from someone else.

Michael Heilperin, of Institute of Higher Studies, Geneva:
Monetary Concept of Capital

Growing dissatisfaction and impatience with current theory of capital. Study of European banking reform.

Discussion between Knight and defenders of Austrian school assumed to be known.

Austrian theory purported to have as object (by Kaldor) to show that capital is a distinct factor, measurable in homogeneous units, and that price is rate of interest, and can be brought in theory on basis with labor and land. [See note 7.]

Investment period concept to give homogeneity.

Distinction between capital goods and natural resources.

Great difficulty of finding what concept covers and how to measure it.

Difference between the theories due to difference of concept, not of explanation. Would suggest still another sense. A serious difficulty, unescapable as long as used for a particular type of goods.

Also difficulty in that various types of production goods considered as capital at different times, as well as trouble of distinguishing production goods from capital goods.

The crucial question is that of measuring a heterogeneous quantity of material goods. Believes is no statistical technique that makes up for absence of common denominator.

Can measure in money, common to all.

Austrian school uses period of production, but requires clear notion of what a period is and how to measure it, and that must prove proportionality between period and the amount of capital.

Henri Poincaré contrasts philosophy and metaphysics with mathematics. Says can agree in latter field because is possibility of verification.[10]

Capital discussion reminds of many philosophical discussions.

Difficult to conceive of period of production as needed for capital theory.

If can't define duration clearly, can't measure as time is measured. Kaldor writes that can use an index to measure variations of period. Heilperin thinks of little help, and can't test exactness of index.

If production period neither meaningful nor measurable, no need to discuss proportionality to capital.

Believes is really a treatment of technical progress, and effects of progress in increasing return to labor.

Grants that technical considerations enter into economic problems, of which another aspect is considered.

Distinction between land and capital historically explained by interest of early writers in rent.

Process of production does include certain elements that are distinct enough to be kept separate: material factors, labor, and capital as a monetary concept.

Money now recognized to be an active factor in the economy, but we retain old, non-monetary vestiges.

First need of entrepreneur is monetary capital, in any productive enterprise.

Must be available for period long enough to enable to produce product and sell it. Time period needed for repayment important.

How money income distributed and whence comes money capital? Saved, but is only what is put aside for use in own investment and for loans. Is also the cash balance, and a third element, held temporarily before disposed of.

Cash balance, consumption, saving and hoarding are the four elements. Keynes should not have disposed of hoarding by neglect. "Saving" may be hoarded again by person holding them for productive use.

Monetary capital is the part of income which is being saved. Is a <u>stream</u>, not a <u>stock</u>.

These funds used in purchase of factors of production in many different markets. Should see effects in each market, their repercussions, and the effects upon investment in the different field. Results may be similar to present – not yet worked out.

But at least puts matter in way more like real world, and releases energies from rather profitless discussion.

Forced saving a queer concept, because saving itself a double act of will (to refrain from spending and to put funds to use). So-called forced saving only explainable by means of price moves.

What of the accumulated stock of equipment? Valuations meaningful only if mean what can sell for; value of total equipment rather meaningless.

Depreciation and maintenance another problem than capital, that of preserving productive capacity. Rate of capital formation related.

Capital can be consumed, if take fund that has been saved and use for consumption. More sense than if think of as failure to maintain.

This concept is a measurable, determinable one. It relates the general theory to that of international trade, which has long so used. Confusion there, thought, because includes also short term funds.

Are consequences upon theories of interest and of liquidity, and upon banking.

Confusion in institutions as in theory. No clear distinction between long-term and short-term funds, even legally (though attempted).

Time deposits not distinguished on any theoretical basis (that which marks off money that can be used for starting productive processes and for other means[)].

Institutions supposed to be distinguished by purchase. Here this type of theory meets the practical problem.

Discussion:

Bernhard: Böhm-Bawerk's theory seems directly traceable to classical wages fund. Is correct interpretation?

H[eilperin]: Does not think was a fund of resources theory.

Jaffé: Want something that will aid in determination of interest. Could have more on relation of this with interest?

H[eilperin]: Thinks interest theory also a composite of explanation of interest and its behavior with attempts to justify it.

Interest is a price for the disposal of purchasing power. Many rates, for different lengths.

Implicitly was always a price, because expressed originally in money. Retained percentage expression even with "real" discussion of capital, which was inconsistent. Rate of interest either determined by market or an influence upon market.

Jaffé: Must go behind money.

H[eilperin]: Capital is power of purchasing the other factors of production. Demand for capital in money terms depends upon prospective returns, etc. Explains inadequacy of purely monetary proposals.

Jaffé: Money concept unnecessary in static economy.

H[eilperin]: Capital theory rather unimportant in static economy.

H[eilperin]: Three factors are capital in monetary firms [sic: terms], followed by labor and material factors.

Deibler: How account for ability of borrowers to pay interest?

H[eilperin]: If this were not possible there would be no problem.

Deibler: Consumption loans?

H[eilperin]: No ability to pay, believes. Critical of consumption credit as an institution, as result of this theory.

Deibler: War loans?

H[eilperin]: Out of capital, or hoardings. Alternative uses would rule price. Might be decapitalization.

Bernhard: Consumption loan made only if think borrower has income whence can liquidate, says Knight.

H[eilperin]: True.

Custis: Once loans were mainly consumption loans.

H[eilperin]: Then considered immoral. Thinks a good case for abolition of interest on short term.

Custis: Marginal productivity?

H[eilperin]: Would prefer marginal rentability, to indicate that productive process has monetary side.

Deibler: How an income from monetary terms? Driven back to material resources.

H[eilperin]: No problem if those did not so work. Operation of system depends upon monetary calculations.

Deibler: When buying material resources and labor, what are paying for?

H[eilperin]: Things that technicians are going to use. Entrepreneur concerns self with monetary profit.

Custis: Is a real productivity with also value productivity.

H[eilperin]: Yes. Processes simultaneously proceeding on two planes. Need not distinguish.

Deibler: Can't answer all questions by monetary analysis.

H[eilperin]: True, and would not attempt to answer all questions thus. Believes the notion of capital more useful as monetary concept. Might eventually do away with capital concept entirely, as ambiguous and leading to more confusion than assistance.

Jaffé: Any interest in a static state? Would have some transfer of goods, replacement, etc., that might give prices for capital goods. Would be a payment for convenience of money being held due to delays, faulty anticipations, etc.?

H[eilperin]: Can start either from non-monetary or from monetary economies. Finds latter more useful. I. All incomes are in money. II. All decisions concern money. Takes process as given and tries to analyze it.

H[eilperin]: On Deibler's remarks. Do not wish to explain everything in monetary terms. Effects of various uses of monetary purchasing power from income determine what is finally done in the field of real goods or technical processes. Economic justification governs, and that is finally a money concept. Thinks that must start from money stream to get to technical processes, though those are basic.

Might have merely a technical problem. The economic problem enters only when becomes matter of valuations in money.

Deibler: Source of monetary capital money savings? What is done with bank credit?

H[eilperin]: Inasmuch as applied to production and not merely circulation, a disturbing factor that stimulates production that can't be maintained.

Bernhard: Why breakdown?

H[eilperin]: Price movements, etc., prevent continuing of financing of capital by bank credit alone.

H[eilperin]: Conception that capital comes entirely from savings represents an equilibrium situation. Otherwise might have excess of savings or excess of funds for productive purposes over savings.

Lachmann: Three factors of production, labor, material resources, and monetary capital.

H[eilperin]: Latter on a different stage.

Lachmann: Is cooperation of all three necessary?

H[eilperin]: Need only two for technical processes. In monetary economy, cooperation of third is needed.

Lachmann: What of the three factors of income? Interest not independent of incomes of material resources. Traditional theory established independently.

H[eilperin]: Challenges this "independence."

Lachmann: Did not mean that were independent of each other, as marginal productivity shows dependence. Have just introduced the entrepreneur.

H[eilperin]: Connected with risk-taking. Combines factors of production.

Lachmann: Risk-taking only in absence of perfect foresight, but other service indispensable. Why not also a factor?

Assumes no perfect foresight.

Deibler: Payment not on a par with that of labor?

H[eilperin]: No. Believe that notion of profit implicit in idea of productive process. Hope of profit, not too often disappointed.

Lochman [presumably a different student]: Entrepreneur bears risk in uncertain world and combines other factors. Refuse to give comparable place to his payment, because erratic.

H[eilperin]: Immaterial to system whether entrepreneur called a factor or not. Adds nothing to knowledge, but adds nothing to confusion, so o.k.

Lochman: Whole distinction of factors of production is terminological. Sought to distinguish incomes earned in some sense from windfalls and rents.

H[eilperin]: All incomes are important to productive process.

Jaffé: Analogous to Walrasian double system, yet relating to each other.

Hohman: Liquidity?

Heilperin: Distinction between his own, the stage at which finished goods exchanged for money and loans repaid, and the ability to exchange an asset for cash because others will buy from one. Believe an important functional difference. System as a whole liquid in first case, but not in second. Liquidity in first case depends upon continuance of economic process. Second idea, marketability, depends upon market situation for particular assets.

Bernhard: Liquid because continue to make loans.

H[eilperin]: True, but because continuance of loans necessary for continuance of economic processes.

Lochman: Would money remain liquid if economic processes stopped?

H[eilperin]: Cash is the standard of liquidity, but not itself considered as liquid.[11]

Dougall: Land, labor, and capital classification as against Heilperin's. If money represents material goods, and goods continuously converted into money in liquid economy, what distinction?

H[eilperin]: Difference is that he includes money, which the classical approach pretty largely neglects.

Deibler: Can reduce to a single factor, money, as all can be so expressed?

H[eilperin]: No, as money used to procure the others, which use.

Jaffé: Must relate to problem, which is how to determine the price of capital and the price of its use. How does this classification help?

H[eilperin]: To his view, important thing is procedure from money decisions to technological actions. Have a supply and demand for money funds, as also for goods themselves. Two separate markets, though related. Money market important only in dynamic world.

Guthman: Actual money markets are rights to land (mortgages about one-third), to government debt (about one-third), and only about one-third to business enterprises.

Deibler and Secrist: These are only the trimmings to the industrial process.

Guthman: Capital actually represents many things not closely related to productive process.

H[eilperin]: Thinks not so far apart. He would separate from concept of purely productive capital or particular type of goods.

January 18

Lochman: Unable to see influence of liquidity preference on interest.

Marginal propensity to consume an important part of his theory and seems to be of considerable validity.

Income in Keynes and in "classical" theory not the same thing. In Keynes it is a problem; in the "classicists" it is not. Latter concerned with distribution of income rather than its size.

[The pages with the foregoing cease or are interrupted (see below), followed by "Draft of my Seminar Report" dated January 18.]

The Propensity to Consume

I. Principal object: similar to "classicists" in using restricted form of a more general consumption function.

   A. Classicists called their's a savings function.

   B. Method essentially the same, ceteris paribus.

      1. Classical: interest; Keynes: income.

      2. Significance of ceteris paribus device – not eliminating the influence of other factors but only of changes in the other factors. [In margin: The essential criticism; for practical purposes.]

II. Will also do two other things:

   A. Consider briefly the form of the Keynesian function

      1. Positive sloping, < 1, for changes in income (and for higher income levels, though no change)

      2. dC_W/dI_W tells how given dY divided between dI and dC. [In margin: Weak]

   B. Indicate its place in his theory

      1. Object to find N. This related to I, which is determined by C, for given I.

      2. Disproof of contention that fall in money wage will raise N by $0 < dC_W/dY_W < 1$.

      3. Element in multiplier, determining effects upon Y of dI.

[The following, found at this point, is typed on a single sheet of paper on which are also several forms of the consumption function stated in terms of these factors. The nature of the stipulated subjective and objective factors, notably in contrast with present-day interpretation, is interesting.]

KEYNES' SUBJECTIVE FACTORS (in the form of motives for saving)

(1) Precaution (reserves against contingencies).
(2) Foresight (provision for expected inadequacies of future income).
(3) Calculation (for interest and appreciation).
(4) Improvement (a gradually increasing standard of living).
(5) Independence (sense of power without definite purpose).
(6) Enterprise (means of carrying out speculative or business projects).
(7) Pride (bequeath or build a fortune).
(8) Avarice (miserliness).

[Added in pencil: "Corp[oration]s: inter[est], liq[uidity], imp[rovement], financial problems]

KEYNES' OBJECTIVE FACTORS:

(1) Changes in the wage-unit.
(2) Changes in the difference between income and net income.
(3) Changes (windfalls) in capital-value, not allowed in calculating net income.
(4) Changes in the rate of time-discounting – in the ratio of exchange between present and future goods.
(5) Changes in fiscal policy.
(6) Changes in expectations of the relation between the present and the future level of income.

[Added below the foregoing, in pencil, are: (1) specification of income in wage units; (2) specifications of General and Particular forms of consumption function to include the foregoing six objective factors plus income in wage units, all modified by eight subjective factors, i.e. the "Particular form for particular levels of objective factors and character of subjective factors."]

[On a separate, smaller sheet is found the following.]

Significance

(1) $0 < dC_W/dY_W < 1$ essential element in Keynes' disproof of contention that fall in money wage will increase employment in manufacture of C.

Expenditures on C rise <u>less</u> than incomes, so total revenue rises less than total outlays of entrepreneurs, and losses result.

(2) Elements in the multiplier. [Standard equations given.] When dC/dY large, k large, and are large fluctuations in employment from small changes in investment. When the <u>average</u> propensity to consume is large, much of the output of society will be consumption goods, and variations in investment will concern a smaller part of the total economy.

(3) For a given investment, Y is determined by the propensity to consume (to save). From Y we may find N, the number of people employed.

[Then follows a draft of Smith's paper, "Keynes' Consumption Function," incorporating the foregoing, and more. Some of the draft is on the reverse side of exams given by Smith in Economics I, dated January 26, 1938. Part of the draft is on the back of the mimeographed pages of a remarkable document, incomplete and with an unknown provenance (though seemingly emanating from the Northwestern University School of Commerce), of some historical interest. Spelling errors have been corrected.]

LOCALIZATION OF INDUSTRY AND LARGE-SCALE PRODUCTION

Professor Gardiner Means, of Columbia University, has made a series of studies on the growth and significance of the large corporation in American economic life. He has discovered that in almost every major industry one or more corporations may be found that have over a billion dollars in assets. He has further discovered that a few corporations tend to control most of the others. Some of his conclusions are as follows: (1) Large corporations have grown between two and three times as fast as all other corporations; (2) the large corporation is coming more and more to be the economic unit with which American economic, social, and political life must deal; (3) an increasing proportion of production is carried on for use and not for sale; (4) a handful of individuals are coming into control of a large part of our entire productive system.

(a) Does this mean that our individual exchange system is being controlled by the few?

(b) Is this one of the criticisms of the individual exchange system made by the socialists, communists, and fascists?

(c) What can you say in favor of such a concentration of control?

(d) What advantages can you see in such a vast amount of control in the hands of a few individuals?

(e) If you were one of the handful of men who controlled most of the productive instruments in industry, what type of organization would you use to retain control – the pool, the trust, the holding company, or the interlocking directorate? Point out the distinct advantages and disadvantages of each of these types of control.

(f) Suppose you adopted the trust form of organization, Would you adopt a vertical or horizontal type of combination?

(g) Take the steel industry. Point out how, because of the organization of very large units, an increasing amount of goods is produced for use and not for sale.

– – – – – – –

(a) The individual exchange system is one in which large freedom is left to the individual both in determining what is produced and how he shall dispose of his income. There are more small concerns than large concerns in most every line of business. In large corporations the man at the head of the business would be the only one who could say how the corporation business was to be dealt with. There are more men determining how small businesses shall be directed than there are men determining the affairs of the large corporations. For this reason and with this understanding of the Individual Exchange System the system is not being controlled by the few.

(b) This could hardly be a criticism of the Individual Exchange System by the socialists, communists, or the fascists. In the socialist order all industry is socially owned, or owned by a few, wherein workers are paid a wage but have nothing to say about operations of the industry. in the communistic order the responsibility for the enterprise would be assumed collectively; hence, by a few. In the fascist order private initiative is recognized but it must be a useful element for the promotion of the state or it will be taken over by the state.

(c) With the concentration of production, or large-scale production, there are many advantages and economies obtained. These are enumerated in (d).

(d) The advantages from such a control are many. Cheap and efficient transportation is developed by an increase in the market. Full advantage can be taken of the gains from the division of labor in a large plant. The larger the corporation the greater will be the opportunity for applying elaborate tools and machines. The large buyer can secure a better price; likewise he can offer a price for the sale of his products. The large plant can afford to install machinery to make use of by-products. A large corporation can afford to experiment with new devices and with new methods. It can also experiment with industrial relations. A large plant can experiment with plans for stabilizing its operations; such as, the forecasting of demand for the product. It can borrow at lower rates and take advantage of all cash discounts. The advantages from large-scale production can be classed into internal and external economies. Internal economies result from the mere growth in size of the producing unit and external economies result from conditions outside the establishment.

(e) A pool is a formal agreement entered into by two or more corporations. The advantage from a pool would be the easy settlement on rates and territories served. The disadvantages of such an agreement would be the limiting of competition. A trust consists of a board of trustees appointed to hold the stock of the corporations to be combined for which trust certificates were issued in exchange for the stocks of the corporations. The advantage of a trust would be that a unified policy of production and operation could be developed without the fear of the effects of competition. The disadvantage of a trust would fall on the stock holders for they would not be able to vote, that being done by the trustees. A holding company owns the controlling stock of the subsidiary companies. The advantage would be that an investor could gain control of a large amount of investments by pyramiding his holdings. The power to control private business would be in the hands of a few who would not have corresponding liabilities. When one company elects members of the board of directors of another company to its own directorate it is known as an interlocking directorate. To me the trust combination seems to be the most sound way to retain control. The liabilities [the document breaks off here.]

[The class notes seem to resume with the following, apparently a seminar presentation by the student Lochman.]

Doubts now as to whether rate of influence has importance in relation to amount saved. Some react in one way, some in others.

Economic theory before Keynes depended mainly upon the downward-sloping demand curve for savings rather than the supply curve.

For Keynes income as a magnitude is variable. Society has unemployed factors. May be result of price and cost rigidities, but these exist.

May be an equilibrium of partial use of resources. If so, how can the use of these resources be increased?

Entrepreneurs do not expand output beyond low level, for don't expect to receive prices that will pay. How to make possible this expansion of output?

(1) Exports, which do not come on local market and depress prices. (2) Investment, which creates employment but does not bring goods back onto market.

Why don't entrepreneurs expect to be able to increase output and get back equivalent of their increased expenditures? Because $dC/dY < 1$. Not all of incomes paid out will come back to them.

Only means of expanding output will be to increase export balance or to increase investment good production.

Latter undertaken on some expectation of future. While investment is going on, society consumes more than it produces.

Extent to which investment is favorable depends upon marginal propensity to consume, for that determines relation of income to investment.

Let dC/dY be 1. $dC/dY = 1 - 1/k$. Then $1 - 1/k$ must be zero, which requires k equal infinity. Would be multiplier of infinity, and increase in investment will cause continuing expansion of output until have full employment, and then inflation.

As to practical applications: Keynes' assumption seems to be of homogeneity for all labor and all other resources, so could combine either with each other. Keynes mentions influence in "bottleneck" discussion. At point where have no longer surpluses of every needed factor the Keynesian analysis breaks down, Lochman feels.

Crane: What of Keynes' approach to present U.S. problem?

L[ochman]: Don't know what "full employment" of all resources means. With rigidities, limited mobility of labor, and specifically limited capacities of capital goods, Keynes' policy inadequate. Dealt with problem in paper in A. E. R. of September. When have full employment in particular branches of industry, would be rising wages and prices in some industries, in others unemployment, and then speculation. Problem seems insolvable.

Keynes, in <u>Trade Cycle</u>, allows that any policy dangerous that raises expectations of entrepreneurs that will be disappointed. L[ochman] believes continued attempt at <u>full</u> employment will thus be dangerous, unless were completely water-tight long-time and short-time money markets and could retain former at proper rates while raised short-time rates or rationed credits to prevent speculation.

Keynes, speaking, has said in times of evident lack of surpluses in <u>all</u> sectors of economy, that the problem is then one of depressed areas, and particular measures should be taken. In this condition the older theory is again applicable. It says that with free enterprise, etc., the needed adjustments will be made but very slowly.

Deibler: Is necessary to develop theory <u>outside</u> of classical approach along lines of Keynes?

L[ochman]: Thinks are really two problems. The one, of depressed areas, is within the classical orbit. The other, of general unemployment, is one in which Keynes' approach is useful. Perhaps classical theory could do the job.

Do not think particularly useful to emphasize similarity of all theories. An approach of different nature and different orientation.

Older economists dealt primarily with distribution of income, but were also concerned (see A. Smith) with additions to wealth.

Discussion over savings equal to investment largely futile and terminological. [See note 1.]

Keynes considers investment as a more or less definite thing, determined by decisions that are more or less given by the psychological attitudes and the relation between the marginal efficiency of capital and the rate of interest. The latter not always effective, he grants.

Deibler: Better as matter of policy to seek to remove rigidities or to meet general unemployment when it occurs.

L[ochman]: Depends on whether believe rigidities can be removed. Moreover, believes that are two different problems. Keynes does not deal with depressed areas.

Keynes considers money wages rigid, not real wages. Has no function connecting real wages and supply of labor.

Secrist: How much of theory balderdash in that uses high degree of conceptualization, neglecting differences of mobility of labor, etc. Any hope of economic science, proceeding along lines of conceptualized scholasticism with no referent for meanings or no possibility of testing with experiments, etc. No theory tells of a policy. That is a question of what we want, who wants it, etc. [See note 7.]

L[ochman]: Now the general question of the methodology of the social sciences. Unable to experiment.

Secrist: We are making experiments and recording results. Raises question of whether can give logical, conceptual picture of way in which economic system works, apart from men's actions and the records of those actions.

Deibler: Is Keynes merely dealing conceptually, without attempts at verification. [See Note 4.]

L[ochman]: Very little of statistics in text. Do not accept statement that Keynes deals with world as ought to be rather than it is. In absence of ability to actually experiment, can only try.

Deibler: Suppose wanted to use the statistical technique as verification, need we a conceptual basis? Is as much of divorce between conceptual basis and statistical method as Secrist thinks? [See Note 10.]

Haensel: Must look at actual life, and how reacts in actual life. Must be very careful not to apply directly any theory.

Deibler: When theorist passes to question of policy must be careful, too.

No one can proceed in statistical work without conceptual background. (Secrist agrees.)

Secrist: Mitchell spoke of difference between a theory arrived at by "looking and seeing" and between pure theory. Secrist believes probably more profitable to build up by experience, as in physics. To understand market, must take it as it is and study it.

L[ochman]: Real trouble is that do not immediately detect causes by looking, and we seek causes. Keynes is a shrewd man, and this book is about a particular situation which he observed. Conditions did not fulfill those of traditional theory. Constructed a new theory to fit these observed conditions.

The most complete study of time series will not provide the causes of the business cycle.

Social sciences can explain <u>why</u> things happen, which natural sciences never can. Social sciences do so by explaining as human motives.

Secrist: But <u>why</u> do humans act as do? No progress in this way.

Haensel: All Keynes' assumptions are directed and chosen for support of his policy. [See Note 7.]

Secrist: And if accept all assumptions of Physiocrats, it is water-tight.

L[ochman]: Theory is sort of consistent framework to keep in understanding world around self. Always neglect something. Keynes set himself a certain problem and selected factors that seemed of importance. Experiment can't proceed in social sciences without the individual being an active participant.

<u>February 8</u>

Jaffé suggests that seminar from here on make a study of Keynes directed at his theory of interest and investment. Suggests pp. 135–254 as basis, and hope for multilateral participation.

Pp. 135–74–85: What is his general theory of interest? What parts do liquidity preference, investment, and hoarding play?

O. Lange, Rate of Interest and Optimum Propensity to Consume, February 1938, <u>Economica</u>.

Fleming, Determination of Rate of Interest, <u>Economica</u>, August 1938.

Lerner, <u>Economic Journal</u>, June 1938, Alternative Formulations of Interest Theory.

Mrs. Robinson, <u>Economic Journal</u>, June 1938, Concept of Hoarding.

Joan Robinson, <u>Introduction to Theory of Employment</u>, pp. 65–85.

<u>February 22</u>

Dr. Freulich: (?)

European Experiments in Protecting Small Competition

Results: Excess capacity, extreme rigidities, high overhead costs resulting from purchases of licenses.

May help particular group during first crisis, but usually harms in course of two or three, etc. Difficulties of cycle enhanced.

Small business men join fascist drive in hopes of state intervention to protect them and to help them approach a monopolistic position. Find that the state interested in armaments and regimentation of heavy industries, not in the small, and difficult to control, businesses.

Practically amounts to a <u>continuation</u> of medieval guild regulations. Almost no period of <u>laissez-faire</u> in Austria, etc. Very little feeling for free enterprise in middle class there. Large industrialists and financiers concerned more with politics, social policy, etc.

Up until two years ago sentiment for free enterprise found mostly among socialists (workers) and bankers. Did not seek complete freedom – wanted strong trade unions and maintenance of tariffs.

<u>March 22</u>
[These notes were so dated but found after those of April 8.]

Lange places essential characteristic of classical theory in assumption that interest-elasticity of demand for liquidity.

Assume rate of interest not affected by changes in quantity of money in circulation, save for short run.

Interest related to quantity of savings (equals investment at equilibrium) and to marginal efficiency of capital.

Deibler: Saving and investment used in same sense here and by Keynes?

Jaffé: Savings same.

Custis: Some would allow for waste.

Deibler: Hoarding included in investment?

How accurate is the Keynesian interpretation of Marshall?

Think not necessarily true that classical theory expressed in money.

Jaffé: Is no "classical theory." Refer to the common pattern of textbooks.

[Diagram with interest rate of return on investment on vertical axis and $ on horizontal axis, with negatively inclined DD curve intersecting unlabeled positively inclined curve at i level.]

D[eibler]: DD represents two things: diminishing prod[uctivity] of added instruments of like type, and also diminishing utility of added like products.

Jaffé: Can't we say that all these forces have a resultant in diminishing return to a dollar's investment.

D[eibler]: When begin to discuss in dollar terms alone, become skeptical, because believe they secondary. Would not the phenomena of balancing of efforts and satisfactions exist anyway? Should not make the measuring stick a primary element.

Jaffé: Balance of satisfactions applies to cash as well.

Custis: Keynes' ideas apply to changing rates, not to high or low rates as such.

Jaffé: If have perfectly static state have no problems of interest, of foresight, of investment, etc.

D[eibler]: Could have a moving static state?

Secrist: If knew the routines according to which changes were occurring.

Jaffé: Would be no uncertainty, but would be investment.

C[ustis]: Moving static state will have an equilibrium rate of interest.

Can interpret "statics" as economics of equilibrium, and no attention to how got there. Older idea is of forces tending to bring about equilibrium. Keynes largely dealing with short-run point of view. Does he really join issue?

Jaffé: Classicists failed to treat of influences upon position of equilibrium of adjustments that are to lead to equilibrium. [See Note 12.]

D[eibler]: Classicists neglected money as an <u>active</u> force. Keynes et al. go too far the other way.

Secrist: All this pure verbalism, isn't it? Refers to <u>the</u> interest, when risk absent. <u>Is</u> no such thing. [See Note 4.]

D[eibler]: In any realm of science a conceptual notion precedes the physical.

C[ustis]: Law of falling bodies in a vacuum is quite unreal, as are a tremendous number of physical concepts. They have their pure conceptions as do we.

Jaffé: Could have an equilibrium for any particular rate. Classical theory places where rate of return for given volume of loans equals volume of savings available for that particular type of loan.

S[ecrist]: Perhaps could verify these concepts for a particular rate, but not for rates as a whole. Perhaps could have some validity as an average, subject to limitations. But not effect of net interest upon total investment – that can't be verified.

Jaffé and D[eibler]: Verification continuous, and brings changing concepts with it.

D[eibler]: Have a structure in advance, which attempt to verify.

S[ecrist]: Begin with observations upon which can agree. Theory is a pattern which try to apply to new cases not falling within the observations. If succeeds, accept tentatively.

Jaffé: Keynes (article in <u>Lessons of Monetary Experience</u>) agrees with Classicists: (1) Interest on money premium for present money in terms of future; (2) Money not alone in having marginal efficiency in terms of itself, a premium for present cash.

D[eibler]: Assume no change in "goodness" of money medium. Such changes introduce another factor [preceding three words seemingly added later] independent of time preferences.

Jaffé: (2) Surplus stock in excess of requirements may have negative marginal efficiency in terms of selves, but usually positive.

(3) Effort to get best advantage sets tendency for capital assets to exchange in proportion to marginal efficiency in terms of some common unit. Net expectations equal.

(4) If demand price for asset A not less than replacement cost, new investment in A will occur.

D[eibler]: This the margin of investment.

Jaffé: What this increase will be depends on: (a) elasticity of supply of instrument; and (b) elasticity of demand for output of this capital good.

These propositions agree with orthodox theory. Marginal efficiencies of all kinds of assets came to be equal, and consequently to equal rate of interest.

Forces determining common value held to be independent of money, which has no autonomous force.

Interest determined by marginal rate of return on capital assets.

S[ecrist]: Essentially a productivity theory.

D[eibler]: Not entirely, for has cost factor, and in this could be included time preference.

Jaffé: Money itself not considered as an investment (having a marginal efficiency of its own). [See Note 1, regarding real vs. portfolio investment.] Keynes' theory holds that in wide range of cases money the most important.

(5) Classical says: Marginal efficiency of money in terms of self is independent of quantity, and in this respect differs from other assets. From strict statement of quantity theory of money. Idea that only effect of quantity is upon prices. Difference between two separate states, each completely achieved. No reaction upon interest rates, for all prices up as much as holdings. But intransition from one level to another have quite different sort of thing.

Based on assumption of full employment, says Keynes, for assume prices will move up proportionally with quantity of money, which requires same quantity of goods, which requires that be no change in employment.

D[eibler]: What is full employment?

C[ustis]: Fisher says T pretty well fixed, labor being willing to work little longer for a time, but not permanently, and will come back to usual amount. Real wages determines quantity of labor offered, and these are same when reach new price level equilibrium from new quantity of money. Keynes says can't lower money wages, but can lower real wages and make possible more employment.

Jaffé: (6) Scale of investment will not reach equilibrium level until elasticity of supply of output as a whole has fallen to zero. (Full employment.)

Synthesis

Equilibrium rate of aggregate investment, corresponding to level of output for further increases in which elasticity of supply is zero [this clause apparently added later], depends on readiness of public to save and this upon rate of interest.[12] Quantity of savings given at each rate and these determine scale of investment.

C[ustis]: In classical theory, savings that are hoarded are not savings.

Jaffé: Scale of investment settles marginal efficiency of capital. Determinant, and rate of interest, and marginal efficiency of capital must be equal, as saving is determined by rate of interest and investment by latter.

Substitute (5*): Marginal efficiency of money in terms of self a function of its quantity.

6*: Investment may reach equilibrium rate <u>before</u> elasticity of supply has fallen to zero.

Marginal efficiency of capital variable with scale of investment and this unknown until elasticity of supply of output as a whole has fallen to zero.

Fisher's equation of exchange postulates full employment, else I would rise and <u>P</u> not proportionally. Higher rate of interest leads to more economical use of active balances. Inactive also affected, except in long period equilibrium and constant expectations.

Open-market operations of banks affect bond prices and yields (rates of interest).

Orthodox theory has no uncertainty or inactive balances for speculative or precautionary use. Refuses to regard the marginal efficiency of money in any other light than other capital assets.

Real issue with regard to existence of rate of net return on money as such. Why? Because money kept for inactive balances as well as active. Is a limit to inactive holdings: how much is worth while to hold – indicates is a marginal net return on money. Holding of inactive balances is an investment. [See Note 1.]

Asman[n] asks what meant by inducement to invest?

Jaffé: Discrepancy between rate of interest and rate of net return on investment.

Haensel: Theory interesting for definite conclusions that can be obtained. Need to believe something that is susceptible of proof.

Interest problem too complicated for theory, perhaps. December 1938 Treasury financing at negative interest, as desired to avoid personal property tax by holding government securities on January 1.

Are hundreds of things opposite to all these conceptions. Must study actual life; can't study logical theory alone, for actual life doesn't follow it. [See Note 4.]

Bonds: Know can buy 3% municipals as good as any governments that are as good to a millionaire as 10% corporates of same quality. Are worth less to others who have to pay different income taxes.

Jaffé: May be dangerous and bad, but see theory as process of gradually approximating reality in effort to get a more comprehensive view. How to expand to gain reality without destroying the structure of relationships.[13]

H[aensel]: Theory important, but when draw such definite conclusions as does Keynes, must observe actual life, which is full of so many things, that perhaps cannot obtain any definite answers. Not so much that is really new in Keynes.

Most important theoretical inventions have been the product of experiment, not of mere logic. We exaggerate the importance of going into such detail with it. Looking into practical life perhaps we can draw more useful conclusions.

D[eibler]: Looking involves classifications based on conceptions, which are results of some theorist observing facts and trying to explain. Keynes has tried to build structure to explain reality. How can study facts as they are without some basis of classification. [See Note 10.]

J[affé]: If demand curve for capital falls, marginal efficiency falls, and is it logical to hold that incomes are the same and that the supply of savings is unchanged?[14]

H[aensel]: Income tax, affecting differently different classes of investors, destroys entirely this story of investment. Changes whole problem of investment.

J[affé]: Only adds another force, which does not overthrow old relations of forces nor invalidate.

S[ecrist]: Right. But hypothesis no good if doesn't explain new facts and be subject to continuous verification and extension.

H[aensel]: Can't use these theories in his field.

D[eibler]: Should try with theories of incidence before condemn it as useless.

H[aensel]: Doesn't think Keynes' ideas entitled to be put in practice, as he hasn't applied to the world.

D[eibler]: Kettering once said motto of scientist should be, "It can be done." Should not say that any line of investigation must be fruitless.

J[affé]: Much of language of Keynes troublesome because based on oral tradition of Cambridge.

April 5

Dr. [Arthur] Schweitzer on Spiethoff's business cycle theory.

Question why he became a student at time when not regarded as regular movement.

As a boy lived through a big crisis in which his father's firm went broke.

Studied economics under Schmoller and Adolf Wagner. Former convinced was only one method for econ[omics] – the inductive. His method to arrive at general rules from statistical investigations.

Influenced by a Frenchman who said must be downswing after every upswing. Another said could not explain general crisis from classical standpoint. Third was Tugan-Baranovsky.

Three points: (1) Standard cycle, can divide into phases and give a theory that will stand for all cycles in history. Main effects assumed to repeat. (2) Typical pattern. (3) Explanation.

(2) Have upswing, downswing, and crisis. Latter merely monetary. First two in every cycle. His typical pattern has groups of facts: causal, associated, and resultant.

Investment, production and consumption, and crisis, in same order.

Investment the main fact. See symptoms in money market, capital market, actual goods, etc.

Impossible to find direct measure of investment. Takes increase in number of established firms during upswing. Can also by [buy?] securities: stocks first, then bonds, and then bills.

Interest rates not in close correspondence with cycle. Short-term controlled by central banks – monopoly price. Long better, but not too good. Increased volume of money by banks borrowing from central bank.

Neglects bank credit – was unknown in theory of his time. Credit conditional, not a causal effect. Not possible to have inflation.

Production and consumption of goods: Producers' goods distinguished from consumers'. All durable goods are producers' in his sense. Consumption of durable goods always increases <u>before</u> production at beginning of cycle. Not production leading demand. After process started, then circular.

Pig iron, stone, etc. the leading durable goods.

At end of upswing decrease in consumption of durable goods leads. Production of leading durable goods not always closely correlated. In Germany and U.S. deviations greater than in England and France. Coal and cement do not lead as pronouncedly everywhere as pig iron.

Relation between production and consumption [of] consumers' goods not clear. These not very closely related to the cycle. No foundation for under-consumption theory. No causal relationship between agricultural production and business cycle. Mostly German figures and not very reliable. Newer investigations show otherwise.

Price movements: Distinguished general level and that for producers' and for consumers' goods. Price rise usually waits for revival, starting after demand up for producers' goods. Not a cause of cycle. Price level for consumers' goods shows no correspondence with cycle.

Concludes that investment the first and main effect in upswing, bringing demand [for] producers' goods, production of them, and then overproduction.

Crisis: (1) Extraordinary speculation and straining credit system. Breakdown financial system. Not inherent, but psychological effect. Mostly concerned with security prices. Must be psychology as well as price rise.

Speculative crisis in either stock or commodity market. Former concerns prices of securities; latter concerns volumes of goods on market. Distinction between earnings of firms and prices of securities. Can go as long as credit available.

2nd kind: New firms with sole purpose of issuing securities taking advantage speculation.

3rd kind: Scarcity of capital. New firms financed by instalments. Break.

4th: Credit. Combination of all. As soon as credit beyond real assets, crisis inevitable. Banks seek help from central bank. If get it, may be crisis in currency, and devaluation.

In extremist form, always currency crisis and central bank devaluates.

Process of hoarding after such crisis. One of first to point out importance of hoarding (1910).

Thinks crisis will disappear. Three periods history: (1) Only speculative crises – Mississippi Bubble – to 19th century. Disappeared by end as banks so strong, etc. Prediction failed.

Downswing: same as upswing, on whole.

Investments, goods, prices.

Here the lead mostly by consumption producers' goods, followed by production, profits, and then fall in investment. If again to be possibilities for investment need: (1) extension of market; (2) techniques of progress. First must be a process of dishoarding to pay for investments. Need no expansion of money as hoarded comes back. Later says need flexible credit system.

Goods mainly consequence decreased demand producers'. In some crises a standstill production and consumption at beginning, for two years, then upswing, noticed first in pig iron and coal production and consumption.

No strong uniformity in consumers' goods. Wheat output sometimes rose during depression.

Overproduction the main question before downswing. Necessary consequence upswing.

Monopolies can remove it for selves by dumping exports. No good for general problem.

Always consequence of upswing. Can't remove but by downswing.

No general overproduction. Concerned always with producers' goods, and not with all of them.

Are shortages of some producers' and consumers' goods. Shortages of complimentary goods, bottlenecks: coal, skilled labor, money capital.

Heavily falling durable goods prices during downturn and depression. Consumers' goods not related, except for drinks (usually imported).

Main achievements the concepts of typical pattern and standard cycle.

Classification of facts may be wrong or insufficient, as think it is.

Mitchell has regular ["referent" written above] cycles and specific cycles. Latter may or may not be correlated.

Spiethoff has close connections in his theory.

Shortcomings visible: No opportunity to treat time series statistically. Very poor statistical technique, by present standards. Figures not very reliable. Facts do not always support his conclusions.

Main concept of typical pattern correct.

Three principal objections: (1) Investigation too limited. Neglects volume-of-income movement and employment movement. Pattern must be extended. (2) Explanation of crisis. Are different crises, in different times. (3) Believed cycle only possible in and limited to capitalism. Fear an invalid prediction. Probably are changes in typical pattern of business cycle.

Explanation: Arrives at conclusion from typical effects.

How induce investment? (1) Compensating force of depression. Dishoarding, after all stocks of goods removed. Interest-bearing bank deposits may be kind of hoarding. Individual hoarding and hoarding for whole economy differ. (2) Technical progress and new markets. These must be available to get new revival. In absence, can have only very short and unimportant upswing. Conditions for new investment: unemployment, idle factories, etc.

Expansion, once started, spreads circularly.

How also a recognized cycle of consumer goods? Can only be due to increase in wages, he says. At certain point of upswing this increase checks expansion producers' goods. Limits profits and necessary savings for capital.

Income: (1) Capitalized – money saved and invested; (2) for consumers' goods. (1) fails at certain point in upswing.

Overproduction maladjustment between savings, investment, and consumption and production of durable goods. Unlimited prod[uction of; possibly "producer"] durable goods the most important: (1) Impossible to estimate demand in long run, as durable. (2) Sombart's distinction between organic and inorganic goods. Argued that could extend industrial production without limitation, but have limited possibilities for agricultural production, and get imbalance between organic and inorganic. Did not always accept this view.

Overproduction of special groups. Scarcity of complementary goods explained from scarcity organic goods. Skilled labor, and goods and agricultural goods.

Scarcity of capital merely an expression for shortage of special goods. Can't take merely as shortage of money capital. Monetary expression of scarce capital in kind.

Explanation of downswing: Concerned mainly with hoarding. Asked especially if could be hoarding on the whole, despite classical theory of scarcity of capital. In some phases of cycle can have surpluses of capital.

Theory of capital quite different from classical.

Has capitalized income, acquisition capital, machinery capital, etc. Thinks possible to treat whole thing from capital point of view. This the only point where indebted to any important economist.

Different facts influence form of downswing: Unequal distribution of income hastens upswing and lowers downswing. Capitalized income that is not used in downswing. If technological progress enables cut of employment on downswing, no aid to whole economy; but less employment. Monopolies can influence, but cannot aid whole economy. Merely change form of overproduction; can't remove it.

DISCUSSION

Jaffé: To what extent do these phenomena take place within economic framework and to what extent within non-economic framework of discoveries, psychology, politics, etc.[15] Are there any laws of technological improvement or of expanding markets?

Custis: Rise in prices held a pure consequence.

S[chweitzer]: Has an effect in generalizing producers' goods expansion to consumers' goods.

Custis: Would disagree with Keynes and with idea that expansion of money would lead to expansion of trade.

S[chweitzer]: Yes.

Bernhard: Theory really one of a disproportionality in economic system. Why no possibility of removing overproduction, if latter only in certain lines.

S[chweitzer]: Disproportion due to difference between organic and inorganic, is natural: Disproportion not due to error.

S[chweitzer]: No doubt that Sombart's explanation wrong, but he couldn't find anything better.

Bernhard: Spoke of hoarding, but not a monetary theory, so must be goods. Which goods?[16]

S[chweitzer]: Says hoards <u>money</u>. This hoarding is expression of lack of use of durable goods, as pig iron, etc. Originating fact held not to be on monetary side, but recognizes that is a monetary economy.

Crane: Recovery came with increased consumption of durable producers' goods without prior increase consumers' goods. Why should come in this way?

S[chweitzer]: Doesn't base work on value theory. Comes from capital theory to investment, etc.

Motive is incentives of profit, and save for this reason.

Crane: Must not first be an increase in demand for consumers' goods?

Deibler: Always produce in <u>anticipation</u> of demand. Studebaker's new car. Bell: Great 19th century expansion.

S[chweitzer]: His technical progress and expanding markets amounted to same thing as above.

Hoarded capital when released spent for durable goods. Capitalized income. Hoarding of capital, not of would-be consumption funds.

S[chweitzer]: Treats interest rate as result, not cause. Supply of money held to be a result. Money available as soon as have opportunities.

S[chweitzer]: Rise in stocks preceded rise in bonds, was said. Perhaps institutional.

Dougall and Wardwell: Bonds lead here, and have. Better-grade bonds lead about six months. Speculative bonds run about with stocks.

S[chweitzer]: Typical pattern needs adjustment to the different countries.

Haensel: Started by study of facts, without previous theory. Baranowsky approached from standpoint of theory first.

Spiethoff found iron and coal industry behaving differently at times of crisis, and had to fit theory to it.

Remarkable that he pointed out rôle of central bank in times of crisis in controlling cycle, in time when was unheard of and politically impossible. Perhaps now should attend more to banking policies and less to money.[17]

Asmann: Would apply to agricultural countries too? In Sweden? In Australia?

S[chweitzer]: Yes. In Sweden, producing ore, often very pronounced.

In Australia, no. Needs much modification.

Jaffé: Support denial that no relation between agricultural production and the business cycle? Agricultural output increasing may be backward rising supply curve.

S[chweitzer]: May be another sort of relationship, yes. Based on rising population.

Wardwell: Does try to explain each cycle by same reasons?

S[chweitzer]: Yes. No distinction between major and minor. In Europe have only major. He has approximately same effects in each, but admits of variations from the standard cycle. In this respect, quite flexible.

April 8

Jaffé on first part of Oskar Lange's article on rate of interest and optimum propensity to consume. An attempt to relate Keynesian and orthodox doctrine in light of a more general doctrine. Problem to relate rate of interest to: (a) liquidity preference; (b) marginal efficiency of investment (Fisher's rate of return over cost); and (c) propensity to consume.

Choose a common denominator of value, because eventually of service in relating to theory of employment. Unit the wage unit, W. $Q/W = M$ (quantity of money in terms of wage unit).

Desire cash balances for: (1) convenience; (2) general security; and (3) speculative (if expect rate of interest to rise, causing bond prices to fall, better to hold cash than bonds).

All money that exists held at any given time by sum of individuals.

(1) $M = L(i,Y)$, a function of rate of interest and income.

Liquidity preference varies inversely with interest rate (cost of holding cash). $Li < 0$. $Ly > 0$. Varies directly with income, if only for size of convenience balance.

[Diagram with interest rate and money on axes and three negatively inclined Y functions – Y_0, Y_1 and Y_2, from left to right. Interest rate i_0 intersects Y_0 at point corresponding to M_0 on horizontal axis.] Y_0 the curve for a given income Y_0, Y_1 and Y_2 successively higher incomes. Set of curves show way in which liquidity preference changes with change of income, if i constant.

Interest regarded as a rate which equates total quantity of money in existence with demand for cash balances. (What of balances held by banks, and central banks? What of creation of credit money that isn't used?)

Propensity to consume and relation to income and interest rates:

Consume more as income rises, but probably not as much more as increase in income. 45 degree line the limit.

(2) [Diagram with C and Y on axes, 45 degree line, and (unlabeled) consumption function starting at origin.]

$$C = \emptyset(Y, \ i)$$

$0 < \emptyset_Y < 1$, $\emptyset$ greater than or equal to or less than 0

General statement of way in which consumption varies with interest rates not available. May save more for a [time] way with a rise of interest rates and then less with further rise.

[Diagram with C and Y on axes and three gradients extending from origin, labeled, from top down, i_0, i_1, and i_2. Arbitrary level of C_0 corresponding to Y_0 through intersection on i_0.] Constructed on assumption consumption falling with rising rate of interest.

Want relation of volume of fresh investment with propensity to consume and rate of interest.

(3) $I = F(i, c)$. Investment varies with interest rate. Must distinguish marginal efficiency of investment from interest rate. Come to be equal in equilibrium. How is well explained by Fisher in his <u>Theory of Interest</u>. Is an equality that defines position of equilibrium but not an identity.[18]

$F_i < 0$. $F_C < 0$. Fresh investments vary directly with consumption. [Diagram with interest rate and investment levels on axes and three negatively inclined functions, C_0, C_1, and next unlabeled (though C_2). i_0 on vertical axis corresponding with I_0, I_1, and I_2 through intersection with three curves.] Slope of curve based on marginal efficiency of investments.

(4) $$Y = C + I.$$

Four equations and five variables. If know any one of the variables, have some of conditions for solution.

(5) $$Q = MW.$$

If have given M_0 and Y_0, can find i_0. If have then Y_0 and i_0, can find C_0. If have then i_0 and C_0, can find I_0. If, in addition, have $I_0 + C_0 = Y_0$, would have equilibrium. If not true, what adjustments necessary to bring to equilibrium? [See Note 12.] $I_0 + C_0 = Y_0'$. Go to different Y curve in (1), etc., holding M_0 constant. Then have $I_0' + c_0'$ not equal to Y_0'. Finally, through adjustments get $I + C = Y$, and equilibrium.

If propensity to consume increases or if capital goods become technologically more efficient curves of (3) would shift upward. For any given rate of interest, rate of investment would increase, and also income, shifting upward and to right curves of I, and rise of rate of interest. As in classical theory, increase in marginal efficiency brings rise in rate of interest.[19]

If reverse (propensity to save rises) curves of (3) fall, I falls, Y falls, curves of (1) fall, and i would fall. Rate falls with increase in propensity to save, verifying traditional.[20]

Deibler: Marginal efficiency and marginal product not same thing. Whole scheme is in money's worth here, trying to eliminate fluctuations in purchasing power of money.

Jaffé: Argument so far has been schematic and structural. Must know something more about these functions, slopes of curves, and their relations. (See Note 12.]

(1) Liquidity preference: Assume $L_i < 0$ and $L_Y > 0$. How fast these responses? Responsiveness of M to changes in interest and in income known as interest – and income – elasticity of demand for liquidity.

What if propensity to consume and marginal efficiency of investment both change? Depends on elasticities. If interest elasticity high, Y [sic?: I] curves of (1) shift greatly with change in Y, due, for example, to increase in marginal product of capital. Savings increased more than proportionately. Rate of interest reacts sharply to change in income if total amount of money fixed.

[Diagram with interest rate and money on axes, with negatively inclined Y curves and inelastic supply of money at M_0. i_1 and i_0 each intersect Y_1 and Y_0 at points on M_0.] High income elasticity of demand. If M_0 fixed, great change in i; if i_0 fixed, M changes greatly.

Leap from i_0 to i_1, the greater the steeper the Y_0 curve at i_0, or the smaller the interest elasticity (read as any demand curve elasticity).

Interest elasticity greater: [Similar diagram with flatter Y curves.] Reaction of i to changes in income is greater the greater the income-elasticity of demand for liquidity and the smaller the interest-elasticity of demand for liquidity.

Custis: Ambiguity about concept of interest rate.

Deibler: Trying to explain interest wholly in monetary terms. If income-elasticity of demand for liquidity is zero, rate of interest not responsive to changes in income. Demand for liquidity changes only when interest rate changes. [Diagram with i and M on axes and negatively inclined curve.] Not a family of curves, but a single one. Income irrelevant. Curve gives interest-elasticity. $M = L(i)$. Amount of money determines rate of interest independent of changes in income, propensity to consume, or new investment. If interest elasticity infinite, liquidity preference not affected by interest rate. This is Keynes' case, says Lange. [Diagram with i and M on axes; flat interest rate and two levels of quantity of money.] Rapid changes in liquidity preference in periods of great uncertainty irrespective of rate of interest.

Mary Wise: This is not Keynes' assumption (that interest elasticity is infinite).

Jaffé: Doubts it. Keynes doesn't make the assumption for all the time.

Deibler: Dealing with aggregates, not the individual reactions. Where are the marginal desires for liquidity, etc.?

Jaffé: Have equilibrium rate of interest when M (desire for cash balances) equals Q (quantity of money). If M > Q, people try to sell securities to get cash, and will be marginal offerers and bidders, as in classical theory, and will establish equality by raising or lowering rate of interest. Effective liquidity preference the marginal preference for each man.

Deibler: An equilibrium theory treating interest as a monetary phenomenon.

Jaffé: Another special case, where interest elasticity of demand for liquidity is zero. [Diagram with i and M on axes and vertical quantity of money.] $M = L(Y)$. People want to hold certain part of income regardless of interest rate. Depends on income, not interest rates. Let $M = kY$. Then $Q = kY_W$, and if quantity of money given, income given, and i determined exclusively by equations (2), (3), and (4) assuming Y constant. Interest determined here by propensity to consume, marginal efficiency of investment and condition that investment equal excess of income over consumption.

$$C = \phi(Y)$$
$$I = F(C)$$
$$Y = C + I$$

This the traditional classical theory.

Remains to be seen whether Keynesian assumption of infinite interest elasticity or classical zero elasticity is more nearly correct. May be more nearly one at one time and the other at another. Must be neither <u>no</u> saving nor <u>no</u> consumption if to be investment and income. Where the optimum between these points?[21] Lange has geometrical device for its indication.

Custis: Closely connected with oversaving or underconsumption theories of crisis and with monetary theories. Depression troubles from an improper balance between consumption and investment.

Can't tell what will happen to expectations as result of manipulations of interest rate, etc. This is very important in practice.

Deibler: On <u>these</u> assumptions can explain these phenomena by monetary approach. Does not exclude possibility of another approach, that may be more fundamental, as the goods approach.

<u>April 19</u>

Haensel: Gave paper before German society on shifting and incidence of taxation. Had to fight against the purely theoretical approaches that were more popular there. Theoretical studies must be relevant to practical problems. Must always have in view the possibility of giving definite answer to the legislator.[22] Theoretical conclusions should be controlled by practice.

Keynes understood necessity of drawing certain practical conclusions. However, they are not related to or controlled by facts and practical problems.

Need theoretical training, but must also have some practical outlook.

What is new in Keynes not very convincing; what convincing not very new. Book not of great value for practical application, though perhaps has too widely been so used already.

Chemical research that is not related to practical things and actual experience of life is not regarded as very valid.

Approach of theorist could be much more fruitful if had practical matters in mind. Need contemplation of life. Don't trust logic too far.

Deibler: Do not begin in the abstract in every science?

Haensel: Yes, but from hints given by observation.

Deibler: Idea must precede progress in experimentation.

Jaffé: Must distinguish between the historical psychology by which discoveries are made and the logic of proof of truth or falsity. Pure science the latter job. The first field quite undeveloped, and comes in many ways. Problem of establishment of truth is one of most important problems. Living in world suffering from prejudices, and these are the propositions that fear to bring to light of truth, to criterion of proof. Nothing more disastrous to development of science and academic institutions than to decry use of reason. Pure science civilizes and keeps humanity; light is as important as bread.

Haensel: Greatest danger of fostering this approach is scholasticism. A pity that refer to events and relate to loss of reason. Peoples of great reasoning achievements have lost their existence.

Also a great danger in historical school of concluding too soon from too few facts.

Economists at fault by disappointing students, giving them too much pure theory. Need more an approach to actual life and showing of theoretical background to give light.

Custis: Theory in setting gun to hit a target? What of Wealth of Nations?

Haensel: These o.k. Related to actual life. Steam engine invented in Russia as soon as in England, by peasants who knew nothing of theory. Not economically sound at time.

Just say don't want to press theory too far.

Deibler: As build science in any field, build by abstract basis, or line of reasoning or organization running through the observations.

Haensel: Get initiative from practical observation. Don't know where is the beginning of thought.

Jaffé: Regard truth as made up as "if then" propositions. Existential propositions can only be stated as probabilities.

Deibler: Don't know where facts will lead without generalization. Approximations, repeatedly made, make for progress.

Deibler: Would raise question on Hicks' new book. He wants to purge of utility, diminishing utility, marginal utility, etc., by substituting scale of preference. How come? Doesn't this imply gradations of utility. Jaffé thinks so.

Jaffé: [indifference-curve diagram, with Y on vertical axis and X on horizontal axis.] Indifference diagram. Think of third plane vertical to paper and adopt convention that the greater preference indicated by the higher ordinate. [Second diagram with I and X on axes and curve emanating from origin.] Draw this third plane to get the hill. Likewise for Y. Surface may come to apex at

M, where have a maximum. [On neither diagram.] Take planes through the hill parallel to ground or paper, project intersections downward, and get contour curves. The higher the number assigned the greater the degree of preference.

Theorists who draw these usually <u>assume</u> that are convex downward. This assumption really implies diminishing utility.

As move down along curve becomes clear that for each decrement of B the increment of A must be increasingly large. [Rough diagram with indifference curve and lines from points thereon perpendicular to axes.] Merely another expression of law of diminishing marginal utility.

Empirical test of whole theory perhaps indicated: Universe of two commodities, or present universe divided into two categories. Indicate rate of exchange of A for B by index $P_{a,b} = 2$. A set of indifference curves always for given individual. Translate all his income into A. [Diagram with indifference map with B on vertical axis and A on horizontal axis, with budget line $q_a q_b$.] External characteristic, in perfect competition, is price: $0q_b = 20q_a$. $0q_b$ would exchange for $0q_a$. Tangent of $L0q_b q_a$ = price. Want to reach highest point on hill, but must follow only this path. The highest point the combination of greatest preference, the point of tangency to the highest indifference curve.

By shifting price lines can derive individual demand schedule for (B) and supply schedule for (A). Can add for all individuals and get market schedules.

Empirical data will give fluctuations in price, demand, supply etc.

Have been some experiments at U. of C. by Thurstone in attempt to build up these indifference.

Advantage of indifference curve shown long ago by Irving Fisher, one of greatest American thinkers: utility curve regards utility as function of quantity of commodity alone; the above permits utility to be function of all quantities.

Deibler: Hicks says must reject all concepts of quantitative utility (marginal utility, and diminishing utility). Says is no known method of measuring utility; no unit.

Hicks says if utility not a quantity, but only an index of scale of preferences, . . . [ellipsis in original]

Deibler remarks that has been a change of definition here.

<u>May 3</u> (Last meeting)

Compare Keynes with Ohlin: <u>Economic Journal</u>, March and June 1937. Some Notes on the Stockholm Theory of Savings and Investment: Alternative Theories of the Rate of Interest by Keynes. Discussion in September 1937. Alternative Formulations of the Rate of Interest: Three Rejoinders.

June 1938: Lerner, Alternative Formulations of the Rate of Interest. Tries to show that are really complementary, and credits Ohlin with some superiorities.

Smith (me) on Ohlin and Keynes.

Mary Wise on Lerner's article.

Bernhard on Myrdal.

Bernhard: Equilibrium Concept as Instrument of Monetary Analysis

Myrdal's system: Attempts to integrate all monetary theory with general economic theory.

Historical view, analyzing and criticizing quantity theory. Wicksell's modification of quantity theory and subsequent processes.

Criterion for monetary theory: His monetary theory and his terms are operational insofar as that is at all possible. Demand theory so formulated that quantities are measurable and operational.

Wicksell's natural rate not of this world, but will it lead to search for observable things?

"Monetary" equilibrium distinguished from equilibrium with merely a numeraire. Latter excludes credit entirely. In our system, money not only a veil, but an integral part of the system not to be taken out of the system.

Myrdal says money a very active influence as long as have credit transactions. As soon as allow any relative prices to change, fact of credit transactions makes money an active factor.

Marginal product of capital: Representation of physical product requires but one factor of production and one product of comparable quality with factor.

If exchange values determined only in pricing process and exchange value productivity of waiting only in such pricing process, may use numeraire. Not yet credit.

Add credit, and exchange value of monetary unit becomes very important. Monetary interest influences relative prices and exchange value productivity. Money rate of interest thus enters into determination of natural rate of Wicksell.

Borrowing and lending makes money an active factor. Influences types of goods produced or used, and thus relative prices.

Myrdal formulates capital in way similar to Frank Knight: the capitalized value of expected income. Doesn't deal in savings and investment much. Rate of profits equals capitalized value of expected net receipts divided by rate of interest. Interest rate equals ratio of [in margin, for insertion here: "discounted value of anticipated"] net earnings and cost of reproduction of existing capital.

Deibler: Discount implies existing rate of interest.

Bernhard: Knight the only one who has solved this matter of circular reasoning.

Not significant for Myrdal's problem, eminently practical. Theory so formulated that can account for monopolistic elements in the system. Includes various rates of expansion of firms, degree of confidence in one's expectations, etc.

Discusses savings and investment, showing that can bring his formulation into agreement with equality upon certain assumptions. Between decisions to save and decisions of others to invest lies whole problem of monetary analysis. Coincidence (ex-post) of saving and investment the result of profits and losses. [See Note 1.]

Jaffé: Hicks' <u>Capital and Value</u> [sic], Chapter 12, resolves controversy between Ohlin and Keynes in way superior to Lerner's. Says no real controversy. Can mean real capital goods or money capital. If latter, and take Keynes's idea of interest, must use numeraire like wage unit. If real capital, numeraire must be money.

[The notes end here.]

****

[At the beginning of the notes for May 3, Smith indicates that he is to lead the seminar in the discussion of Ohlin and Keynes. Smith's materials include: (1) seven pages of notes entitled "Ohlin vs. Keynes," which presumably was the basis of his presentation; (2) a one-page "Outline;" (3) a one-page summary of Ohlin; and (4) three pages of notes on Lerner's analysis. These are reproduced below, with some editing.]

(1) *"Ohlin vs. Keynes"*

Introduction. Discussion advantages to Ohlin: Initial word, in which set broader problem: <u>Real</u> issue is period analysis. Last rebuttal, and a more general theory of interest (which includes Keynes').

I. Each studies processes of expansion and contraction; each adopts the monetary terms and drops the fundamental assumption of ordinary price and distribution theory – "monetary stability" (national income or MV constant, etc.). Each found interest theory but a part of his problem. Ohlin found relatively little in the writers and apparently had to make some [sic; omission?] for the occasion. Keynes found the first novelty of his theory the insistence that Y, not i, ensured equality of S and I. Arguments "independent" of rate of interest and had to find something else, equalizing attractions of holding loan and idle cash. Says merely states what rate of interest is, following arithmetic and preparatory schools. In similar way Ohlin merely states, following financial and business journals. A more general concept.

II. Ohlin's <u>integral</u> part of general analysis of expansion and contraction: <u>Vide</u> three points of interest theory:

(a) Analysis of markets (exist) for claims and other assets, where prices and thus interest rates determined (as any commodity).
(b) Explanation of kinds of processes resulting from movements of certain interest rates (re expansion, etc.)
(c) Account of connection between these expansion processes and way they occur and transactions on the credit market. Ability (post) to make financial investments as well as willingness (ante). [See Note 1.]

III. Keynes' system has appearance of determining outside general price system. Is an equilibrium analysis and thereby the more conventional.[23] $Y = f(I)$. Equilibrium of quantity of money, propensity to consume, marginal efficiency of capital, and liquidity preference. These fix i, I, and Y and N. Cash given too exclusive a place (defense below) by liquidity preference theory. Interest rate determined in re cash (outside price system) and governs the rest. Keynes assumes other three independent unaffected by change in interest rate.

Ohlin finds the most fundamental objection due to lack of distinction between post and ante. [See Note 1.] Propensity to consume gives relation between level of income and consumption spending, $Y(1-k) = I$ but only true ex-post. [In margin: "Does not show determination."] True of every post period, how little stable. Not true, save by chance, of anticipations. As result of differences between planned investment and planned saving find unintentional savings and investment ex-post, and process of expansion or contraction sets in. Probable that many elements in price system will be affected. Depend on speed of reaction. No tendency toward stable position. Better for cycle. (Period analysis good.)

IV. Ohlin's period analysis: Careful distinction between ex-post and ex-ante concepts. Saving and investment equal ex-post, but not ex-ante. His example: Decide to save $10 million less and to consume $10 million more. To invest same, i.e. investment same. Retail goods merchants' incomes up $10 million. Prices rose, so have $3 million unexpected income and stocks (investment) down $7 million (cost prices). Thus savings only went down as investment, ex-post. In next period perhaps a slightly increased consumption (retailers) and increased investment (reduced stocks). Expansion on.

– – – –

Demonstration of significance of ex-ante analysis: Keynes makes Ohlin's net demand and supply credit equal investment and saving (by use of ex-post interpretation). Interest can't equalize for are always equal, by definition.

Answer: Such equality ex-post for every transaction but doesn't invalidate use of schedules (ex-ante). Indicate alternative purchase and sale plans. Are causal.

Real flaw of interest rate as regulator of saving equal to investment is: Any interest rate possible regardless of saving or investment. Interest does not equalize plans, and many [are] possible, adjustments coming in unintentional savings, etc. No unintentional credit (?). With credit, a given willingness to increase holdings of claims (supply curve of credit), etc. for demand can result in only higher interest rate in a free market. Moreover, are credit markets and credit prices, though not for saving.

This analysis includes Keynes' equality advantages of cash from deferred claims. Relative price of types of claims give interest rate and are equalized, varying with willingness to hold the types and effects of investment and saving upon demand and supply of claims. Rate of exchange bonds and cash determined demand and supply of both. Claims just as fundamental as cash and provide direct link with expected saving and expected investment and whole economic process.

Have skipped Keynes on "finance" before investment made. Supposed to handle ex-ante, but not analogous. Says finance credit revalues; Ohlin's used but once. No.

(2) *Outline*

I. Each an orphan theory of interest

       Expansion vs. contraction
       Monetary terms
       Monetary stability
       Orphan theories

II. Ohlin's integral
       Three parts:

           (1) analysis markets
           (2) explanation results interest rate change
           (3) connection expansion and markets of (1)

III. Keynes':
       (A) appear outside.

           Money and liquidity fix interest rate
           Interest rate fixes investment fixes employment
           Cash too inclusive. Assumes changes in interest rate will not

affect <u>other</u> three (?) independent variables of equilibrium (propensity to consume, marginal efficiency of capital and quantity of money, liquidity). No tendency stability.

(B)   Difficulty doesn't distinguish post and ante: Y(1-k) = I only true ex-post and doesn't show <u>causation</u>. <u>Many</u> price elements affected. Spread of reaction.

IV. Ohlin: period
       Example
       Significance

(3) *"Sum Ohlin"*

(1) Regardless of saving and investment, any interest rate compatible with supply equal to demand <u>ex-post</u>. <u>Whole economic process</u> adapts, including income.
(2) Rate of interest price of credit and governed by supply and demand curves as commodity price.
(3) These curves closely related to willingness and ability to save and invest (and thus interest rate and whole economic process) but also influenced by desires to vary cash holdings, or make financial investments in old assets and by changes in bank's credit policy.

<u>End</u>

(4) *[On Lerner]*

I. On Lerner's analysis
    A. Narrower problem. Basic difference
       Takes up only interest; Ohlin the whole process, in which interest rates but one class of things.
       Basic difference.

[The foregoing – all of "A" – X'd out by Smith but possibly reinstated with "O.K." In corner, also X'd out, to which "O.K." may apply alone: "True of money. As true of Federal Baby Bonds; or Chicago and Northwestern Railway Bonds."]

B. Deficient treatment
    (1) Claims to have given Ohlin's theory without use of distinctions <u>ex-post</u> and <u>ex-ante</u>. Therefore <u>fails</u> to portray accurately. Diagram adds a curve which is only true <u>ex-post</u> to curves which are true <u>ex-ante</u> (liquidity and money).[24] Shows eventually that adding and

subtracting same thing doesn't change. Refuses to consider Ohlin's objection.

(2) Application to Ohlin based on interpretation of "credit" (Lerner, p. 221; p. 223 better) which: (a) includes only saving and "hoarding"; and (b) misreads nature of connection between saving and investment and schedules re credit. Supply and demand credit refers to expected saving and investment. Moreover:

    (a) <u>Supply</u> of credit may differ from saving by reductions in investment (reinvestment) in favor of credit expansion, as well as by increases and decreases in hoarding and by bank policy (financial side). <u>Demand</u> for credit may differ from investment by amount sought to cover expected losses (perhaps ok to call investment for Lerner), to spend on consumption, and, on financial side, by <u>financial</u> investment or disinvestment (buying or selling of old assets). Perhaps not fair add financial.

    (b) Ohlin's connection between investment and saving and credit is through effects of processes of expansion and contraction on whole system. <u>The basic difference</u>. [In margin: "Lerner has not <u>proved</u> all his points; he has not sought to treat whole problem."] [The third page has a rough copy of Lerner's Figure 2 and brief, rough notes comparing Lerner, Ohlin and Keynes on *ex ante* and *ex post* functions.]

NOTES

1. The failure to distinguish the *ex ante* from the *ex post* relation of savings to investment persisted well into the post-World War Two period, though Gunnar Myrdal had made the distinction a decade before Jaffé's seminar. Keynes's argument was that although savings and investment need not be equal *ex ante*, any inequality would lead to a change in income and thereby to a change in the level of savings relative to that of investment, which might itself change. Equilibrium, analytically, could come about only when savings and investment were equal *ex post*. The critical variable of adjustment was income itself, not the interest rate. The only recorded use of the distinction in the notes is in the penultimate paragraph, in a discussion of Myrdal, the originator of the distinction. The distinction is also used in Smith's notes prepared for his presentation on Ohlin vs. Keynes, provided below after the main body of notes.

Another problem was the failure to distinguish real from portfolio investment, only the former constituting spending in the gross national product sense; the latter representing leaks from the spending stream. Still another problem, not entirely terminological, was the ambiguity of what comprised a monetary explanation, for example, in regard to savings and the relation (if any) of changes in liquidity preference to changes in saving. Also, the term "capital" was used in several different senses, whose relationships (where present) were not always clearly drawn, and whose different uses tended to blend with each other. The notes below indicate an awareness that the introduction of credit money, and therefore time, was a vastly complicating matter.

2. This relationship was not always recognized in later macroeconomics, though the basis of the idea of the supermultiplier (on the Keynesian cross, the investment function rises – is not parallel – in relation to the consumption function).

3. The "already" is important; this is not the marginal efficiency of capital, the expected return on new investment.

4. See the discussions, in the notes commenting on Smith's record of Jaffé's lectures on Marshallian economics in Volume 17 of this annual, on the distinctions between purely conceptual and empirical levels of analysis and between conceptual tools and definition of reality. One aspect of this distinction which arises here is the tendency to define the purely conceptual model to be problem-free. Another aspect concerns the ways and the degree to which the purely conceptual model does and does not approximate the actual economy.

5. Eventually it would become clear that the operative point is the relative strength of two forces, employment increasing due to lower wage rates and employment decreasing due to lower effective aggregate demand.

6. Later, New Classical Economics would hold that only the former – voluntary unemployment – would exist, denying, disregarding or minimizing the latter.

7. Several points are relevant: (1) The definition of the rate of interest, like many other definitions in economics – such as "inflation" – have long tended to embody specific theories of the subject, and constitutes a good example of theory-laden fact/description. (2) The use of the term "natural" is a rhetorical stratagem by use of which to establish a (perhaps unintentional or non-cognitive) privileged analytical, interpretive, and/or policy position. (3) Two quite different types of interest theory were forthcoming: monetary and non-monetary, with varying formulations of each, with each specific theory posing and answering different questions, and with the answers being different. (4) The theories of interest held by individual economists often were an amalgam of different specific theories. (5) Theories of interest were also combined with other theories – for example, Banking School, Currency School, quantity theory, Keynes's theory, Hayek's theory – and in varying ways. (5) Like so many other economic theories, theories of interest were generated and/or espoused largely on non- or pseudo-empirical and a priori grounds. (6) Theories are sometimes, perhaps often but by no means always, generated and/or espoused (even if non-cognitively) because they contribute to the reinforcement or advancement of a particular paradigmatic, ideological and/or political point of view. For example, different theories of macroeconomics have varying coefficients of attractiveness, depending in each case on point of view, including, for example, their potential role in the mobilization of political psychology and action. That is, theories have their political coefficients of attractiveness and meaningfulness.

8. The remarkable oddity here is that saving and investment seem to be treated as alternatives in the Keynesian system, at full employment.

9. Two issues arise here: (1) the categories of "profitable" and "unprofitable" investment/spending are neither given nor self-subsistent but a function of circumstances and reactions (for example, impact on expectations) and (2) the utilitarian test of profitability, both per se and without regard to the question, profitable for whom?, i.e. whose gains and costs are included/excluded.

10. Modern philosophy of science and of mathematics and epistemology seriously question either testing by verification or the probative meaning thereof in mathematics. In addition (for example), testing depends on definition, which here in capital theory is the point at issue. It should go without saying that the fact that relations can be expressed mathematically (and subsequent deductive manipulations made) does not necessary mean that the posited relations are true (descriptive accuracy and/or correct explanation). Also relevant here is the theory-ladenness of facts.

11. This is surely odd.

12. Note that here and elsewhere the classical theories of saving – as a function of interest rates – and of interest rates – as a function of savings supply and investment demand – and not Keynes's theories (saving as a function of marginal propensity to save and interest as a function of liquidity preference and quantity of money) are being used.

13. Yet the structure of relationships is often precisely the point at issue.

14. This suggests the difficulty of ascertaining when a conclusion is a matter of logic, within a model, and when a matter of the model itself.

15. The distinction between economic and non-economic, especially rampant in neoclassical economics, relates in part to the construction of the pure abstract model of "the economy."

16. The Keynesian would transform the money/goods dichotomy to a money/spending-and-income flow dichotomy, and absorb the former in the latter. Largely omitted from the discussion are the factors and forces generating and/or limiting spending on new goods and services.

17. Confusion or ambiguity exists as to the meaning of "money" in this context.

18. The reference to Fisher's work, which long antedated Keynes's *General Theory*, is interesting. Note the attention to the technical conditions of equilibrium rather than, for example, the factors and forces which drive the consumption function and the marginal efficiency of investment. One wonders how much the attraction is the mathematical mode of expression – formalism – and how much certain technical issues which just happen to be neatly expressible mathematically; whatever the relative attraction, the two reinforce each other. Also relevant, of course, is the analogy with nineteenth century physics, as stressed by Philip Mirowski. See also Note 1.

19. But within different models and, therefore, for different reasons, or through different routes.

20. See Note 13; also, a particular notion of verification.

21. Thus to the question of the conditions of equilibrium is added the question of optimality of equilibrium – on the road to working out the neoclassical research protocol of reaching unique determinate optimal equilibrium results.

22. This attitude is one source of the neoclassical research protocol seeking putatively unique optimal results. The attitude tends to characterize economists of all schools.

23. The statement that equilibrium analysis is "conventional" is to be noted.

24. This is comparable to a criticism of IS-LM made thirty years later by John R. Hicks, that in equilibrium past uncertainty has dissolved.

VICTOR E. SMITH'S NOTES FROM UNIVERSITY OF CAMBRIDGE LECTURES, 1954–1955

Edited by Warren J. Samuels

Victor Smith spent the academic year 1954–1955 at the University of Cambridge on leave from Michigan State University. His files include notes taken by him from several lectures given in seminars and from a set of lectures on Problems of Econometrics, given by Michael James Farrell, and two lectures on Mathematical Economics, given by Richard M. Goodwin.

These notes are recorded below with the following important qualifications: diagrams are described, not reproduced; most of the mathematics notation is omitted; and only certain non-mathematical statements from the Mathematical Economics lectures are recorded. The intention is to focus on ideas and orientation. Besides, it is not possible to fully present the diagrams and the mathematical notations, and neither are necessary to understand the ideas – though it is important to appreciate that ideas/concepts/theories were being increasingly reduced to and/or transcribed into the language of formalist mathematics.

Only minor stylistic corrections have been made. The comments in the notes are the editor's.

The lectures/sets of lectures are as follows:

(1) Nicholas Kaldor, "Value and Distribution"
(2) John R. Hicks, "Marshall Lectures"
    (a) "Foundations of Welfare Economics
    (b) "Another Shot at Welfare Economics"

Documents on Modern History of Economic Thought, Volume 21-C, pages 111–153.
© 2003 Published by Elsevier Science Ltd.
ISBN: 0-7623-0998-9

(3) Lawrence Klein, "Personal Savings"
(4) Richard Goodwin, "Dynamic Economics"
(5) Michael James Farrell, "Problems in Econometrics"

Smith used the abbreviation "a/c" numerous times. Some uses seem compatible with "on account of" and others with "according to." At least one might read "accumulation of capital" and another possibly "average propensity to consume." The "a/c" usage has been retained.

Several points: (1) The notes record what Smith considered important, not necessarily what the speaker said or emphasized. (2) Seminar presentations are often trial runs on work in progress and, therefore, are not necessarily indicative of final positions, but are suggestive of the process of working out ideas.

The notes – or the lectures recorded in the notes – are of interest in several regards, including: the epistemological foundations of economic work; the manner in which conceptual tools are understood and used, notably the attitude or orientation taken with regard to economic dynamics and econometrics; the manner in which the actual economy is understood and treated; the numerous relationships between some notion of a pure conceptual economy and the actual economy; the extent to which the requirements of technique drive economic theory; the extent to which considerations of applied business practice influence the construction of techniques; and the quest for unique determinate optimal equilibrium solutions.

NICHOLAS KALDOR, "VALUE AND DISTRIBUTION," OCTOBER 27, 1954

(Speaks often in low voice. Hard to hear at back of room (of moderate size).)
Historical approach. Speaking of Adam Smith.

Money cost of production theory is not a theory of value. Merely says one value depends on another. Circular reasoning.

Ricardo made brilliant attempt to rescue Smith's theory from its narrow scope (primitive society). Ricardo the first economist to produce a consistent model of general economic equilibrium.

Key to Ricardo is in Preface to his Principles: problem of distribution (among landowners, laborers and capitalists) is principal problem. Value not his real interest. Was really interested in effects on distribution of particular acts of public policy.

Had to clear lot of things out of way: value problem, for instance. As a theory of value in exchange, his qualifications important. But not really relevant to his main purpose – effects of policy on distribution.

Value: neither existence of profits nor payment of rent will affect propor-tion-of-labor embodied principle of value in competitive economy.

Profit to capitalist a/c wages <u>advanced</u> to labor. Production takes time in sense that consumable products come <u>after</u> today's labor

Two forms of advancing wages: (1) Circulating capital – normally one year – the agricultural cycle, yielding product <u>once a year</u>. (But average period of advance really only six months. But capitalist has to hold harvest until wages paid, so not recovered for one year from date of harvest.) Embodied labor, fruits of which have matured.

(2) Fixed capital. Embodied labor – intermediate products. Not regarded as so important. Exactly contrary to modern view that fixed capital the important part. Think difference not so much difference in technology. Mill, Marx and Ricardo all agreed on importance of circulating capital. Marx: only circulating capital creates surplus value. Mill: wages fund determines wages. As if capital consisted of two kinds of goods, wage goods and equipment. Not true that are two kinds of <u>goods</u>. <u>Are</u> two kinds of capital, circulating and fixed, differing in degree. At a <u>moment</u>, in <u>stock</u> aspect, circulating capital consists of unripe goods.

From point of view of wage fund or subsistence, difference simply that only in flow aspect can capital be viewed as wages fund. Stock and flow aspects differ. The longer the period, the greater the difference between the stock and the flow. All capital is turned over, some more slowly and some more quickly. Continuous cycle, except that in agriculture a periodicity.

As capital is turned over, renewed, it is paid out in wages. Equally true of fixed capital, which is renewed more slowly. May give rise to smaller wages flow than would think, a/c turned over so slowly.

Jevons and Böhm-Bawerk developed fundamental Ricardian view. Implies: (1) production period – time between input and output, (a) passage of mate-rials from input to output and (b) durable equipment.

(2) Capital can be regarded as either subsistence fund or intermediate products.

<u>Flow</u> aspect important as subsistence fund.

To understand classical view must fasten on one aspect: wages paid out of capital. Today we don't regard as too important.

Can't raise wages except as result of saving. Spending invariably bad. Wages depend on what rich <u>don't</u> spend. Mill: The demand for commodities is not a demand for labor. Demand labor by accumulating a wages fund.

Aspect of wages fund which is true is: flow of consumable products is confined by productive activities of society.

Doesn't worry us too much today. a/c (1) can increase flow pretty rapidly if start from underemployment, (2) <u>all</u> consumption comes from this flow (wage-earners may not take large part of it), (3) year to year increase occurs.

Back to values: profits tend to proportionality to wages, among industries. Whether profits high or low doesn't matter.

	Commodity	
	A	B
Quantity of labor	100	50
Wage advance	£100	£50
Profit rate	£10	£5
Total cost	£110	£55
Exchange ratio	2	1

Exchange ratio not altered by level of profit rate. Changes in profit and wage shares leave value unaffected. (But if wages not advanced for the same period, exchange ratio thrown off. Now a change in rate of profit _alters_ exchange rate.) Exchange no longer: (1) depends on quantities of labor; (2) independent of distribution of produce.

Not much worried even though now things affecting distribution also affected exchange value, for held that in agriculture these defects were not important, and that must look to agriculture to see consequences of policies.

John R. Hicks, "Marshall Lectures"

February 24, 1955

Introduction by [Claude] Guillebaud:

Switched from Mathematics to Economics. Oxford undergraduate. Taught at London School, Cambridge, and Manchester, and now at Oxford. Holds Drummond Chair of Political Economy, the oldest chair in England.

Lecture:
Paper on Foundations of Welfare Economics some fifteen years ago. Hoped would bring quiet, but the reverse. No longer desire to defend that position.

Wish to rebuild from bottom.

Theory of welfare is _not_ about welfare. Pigou's book was once called "Wealth and Welfare," and that a better title. The welfare (utility) approach to the theory of wealth.

Theory of wealth: Inquiry into Nature and Causes of Wealth of Nations. Adam Smith's title excellent.

Theory of wealth only a part of economics. May deny that concept of wealth of _nations_ has meaning. If so, not a valid part of economics.

Theory of exchange remains. Economics becomes catallactics. But is that the whole of economic theory?

Schumpeter's History of Economic Analysis always judges economists by their contributions to catallactics. He had less favor for theory of wealth.

Production and distribution of wealth in general goes beyond catallactics. In practice, all economists work with aggregates. They may agree that aggregates require study and criticism.

Here confine to flow analysis.

. . . measure of flow of output. But what weights. Theory of value, as concerned with these weights, is different from theory of price, and belongs to theory of wealth.

Price determined by both utility and cost in catallactics, but in theory of wealth these two approaches provide different meanings.

Begin with cost side. Start with Ricardian model. Let cost measure be [function of] labor embodied. Say social product increased if [flow of labor embodied increases].

Labor needed to produce [larger quantity] at old labor coefficients.

May also get larger product by reducing labor coefficients, . . . Labor theory of value a living element in first chapter of theory of value.

Diagram, not limited to two commodities. [Diagram with number of B bundles on vertical axis and number of A bundles on horizontal axis. Two parallel negatively inclined lines.]

Divide set [quantity] into n units.

In initial situation combine units in A proportions.

Divide B situation commodities into n bundles. Any other points on diagram can be made from A and B bundles put together. [A certain] number of B bundles can be produced with labor available at A.

[Can identify the] relative increase in B social product, using A costs as weights.

Ratios will be same whether based on A or B weights only if lines parallel.

So far, theory has only partially overcome non-measurability of aggregate of commodities by assuming labor measurable. But is a way out. Ratio . . . could equally well be interpreted as relative increase of quantity of B bundles over number could have been produced in A situation. Same principle can be used if factors are many, as long as is same complex at [two points].

But lines may not be straight, a/c changing marginal costs. Increasing marginal cost gives outward convexity.

[Diagram same as previous one, but with parallel convex lines.]

Lines may assume odd positions, but in the real cases, with many products and small changes, would expect only modest bending. Are other cases of importance.

Causes of wealth of nations:

Analysis implies something. Must distinguish between increase in quantity of factors and increase of efficiency.

Increase in labor, with no change in efficiency, moves lines with no change in slope.

Marginal cost, in all factors, if proportionate: lines will at least move in same direction and retain relative positions.

If A and B tests give opposite results, must say evidence inconclusive.

Efficiency increases: With many factors, allocation improvement may increase output from same factors. Then curves have to be defined with and without the fault of organization, and fault applies to whole curve. Improvement must apply to whole curve.

If no further improvement possible, . . . optimal . . . lies within the frontier. Optimal conditions well known.

Ricardo went well beyond simple labor theory. Comparative costs is theory of choice among different types of labor. Rent is theory of cooperating factors.

How regard these optimal conditions? Nothing morally imperative about them.

Must think of a frontier continually moving outward a/c accumulation of technical knowledge and capital. May admit that main cause of increase of wealth is movement of frontier, but width of gap is a problem of importance and the proper task of economist.

Now to statics. Carryover of capital. Whole stock of capital at beginning of period is part of factors. Whole stock of capital at end is part of product.

But raises as many problems as solves. Opening stock highly specialized, restricting range of alternatives considerably. Deprives analysis of most of its interest.

Only lack of full employment liable to keep A below frontier. High specialization may cross. [Reproduces earlier diagram except that now B line crosses A line from above.] No clear answer to problem of whether have output increase.

But in long period, specificity disappears, and theory of wealth may appear better in long period.

But other difficulties spoil it. Time significant. Date of receipt of bread is important. Again unlikely that a reorganization of production will increase number of A bundles. As far as can go by this approach.

John R. Hicks, "Another Shot at Welfare Economics," March 3, 1955

Utility approach:

Limitations of cost approach: [Diagram with B on vertical axis and A on horizontal axis, and with negatively inclined line.] A and B collections differ only in that one of goods in A collection replaced by good of greater (lesser) quality, but same cost. Cannot show improvement (loss) on cost basis. If can't produce B bundles in A situation. In terms of A bundles, increase is zero. In terms of B bundles, increase is infinite.

Interpersonal comparison bogey: Utility <u>not</u> a psychological theory. Simply a means of taking a/c of quality.

Mustn't identify preference machines with real persons. Tastes can't remain unchanged from year to year.

Utility hypothesis a tool for discovery, not a discovery itself.[1]

Settle for rough treatment. Neglect changes in distribution of product.

Suppose N identical consumers with equal spending power. Each receives 1/N of each commodity. Say total product rises if representative consumer bene-fits. Conventional standard of reference. [Diagram with B on vertical axis and A on horizontal axis. Line downward sloping, with several indifference curves; another downward sloping line starting just above first, with indifference curve above it starting at same point on vertical axis.] 0A and 0B represent bundles available to representative consumer. Aa shows positions open at A prices; A is preferred point. Must touch indifference curve between A and B, and, of standard form, lying to right of A except at A. Similarly B.

Natural to think of gain as gain in utility. But utility measure can hardly be better in one direction than in another. From A to B, ratio in B bundles; B to A, ratio in A bundles.

With identical consumers, curves belong to same system, and cannot inter-sect. Measures show same direction of change. Property doesn't necessarily hold for [all changes].

Meaning of Ob bundles: number of A can buy with B income at B prices. But will choose B bundles, not A. [Shows] measure in terms of A bundles of gain from being able to buy B bundles instead of A bundles.

Different individual, so commodities not consumed in same proportions. Diagram refers to market as a whole, not any individual.

If b outside A: money income at B could be divided to permit purchase of A bundles at B prices. But can't conclude could redistribute so as to make each one better off. Effects of change on prices.

[Shows] <u>minimum</u> number of B bundles that can be distributed so as to keep each on indifference curve level had at A.

Collective indifference curves can be generated by using new bundles containing varying proportions of A and B bundles.

Difficulty: Collective indifference curves may intersect? [Diagram with B on vertical axis and A on horizontal axis. Parallel indifference curves plus a third intersecting one at point E.] Only concerned with intersections on the diagram, ... E can lie on both curves only if distributed in different ways as think of on one curve or on other. Must be different incomes, if marginal utility ratios equal price ratios. If are several curves passing through each point, [with] different distribution. What kind?

Change of distribution affects slope on curve at B: Relative marginal valuation of B must be raised if to swing to right. Must be shift in demand from A bundles toward B.

Perverse position of curves requires

(1) large change in distribution of income
(2) large change in demand in particular way
(3) small elasticity of substitution between A and B

Identical consumers:

Laspeyre: [>] indifference measure
Paasche: [<] indifference measure

True measures along indifference curves. Two of them.

Utility measure <u>may</u> fail to stand up to radical changes in income distribution.

Causes of wealth:
Utility optimum must always lie on cost frontier, but need not be just any position on cost frontier.

Compensation principle, a kind of partial analysis. May be useful in discovering possible improvements.

Position at B, on cost frontier, but not optimal, <u>may</u> be made optimal by altering distribution. Not necessarily possible.

May introduce efforts and sacrifices into utility theory. Not usual way to think.

Dynamic interpretation: Let plan B be reorganization expected to settle down after period of uncertainty. Is it enough that <u>then</u> B should be superior to A? But tests difficult. To require to pass them may lead to stagnation.

LAWRENCE KLEIN, "PERSONAL SAVINGS," MARCH 16, 1955

Trying to get at roots of consumption function and put on firm empirical basis.
What is needed:
Don't think simple tricks will help. Need to be highly multivariate, especially at level of individual.

Think of income as being largely beyond control. Simplifying. Assume that amount by which one can vary income in short run (one year) is about two percentage points.

Individual choice:
    Saving:
        Contractual.
        Discretionary.
        Expenditures on durable commodities.

Non-durable Expenditures

Among durables people make deliberate choices. Deliberate commitments for contractual saving. Discretionary: changes in bank balances, government bond holdings, retained earnings, house purchase, alterations in house, etc. Somewhere in this class is a passive element, a buffer that adjusts income and other decisions.

Types of Savers:

    Farmers [and] Businessmen: Possibly all self-employed, for some purposes.
    Others: Low savers (both marginally and on average).
    Farmers high savers (both marginally and on average).

Farmers and businessmen have direct outlets for savings. Is some problem of obtaining access to capital market for these two. Another reason may be variability of incomes. I reject that reason. When their incomes are changing no faster than others, still have high savings. When take their invested earnings out of incomes, find residual behaves much like Others. The English data don't work out so neatly here, as far as have gone.

Variables influencing saving:

    Economic: Income, Wealth, Rate of change of Income
    Demographic: Race, Sex, Marital, Age, Family Size
    Mixed Economic and Demographic: Home Ownership and Occupation
    Psychological (Attitudinal): Pessimism, Intentions, Price Expectations
    Psychological measures rather arbitrary and simple.

[Three diagrams. (1) Discretionary saving on vertical axis and income on horizontal axis, illustrating negative, zero and positive discretionary saving as income increases (positively inclined curve, unlabeled). (2) Unlabeled diagram with three positively inclined curves. "Liquid wealth is good indicator of total wealth; is really what <u>can</u> spend out of capital. Is measurable. Function shifts as liquid wealth increases. Curve tends to fall <u>and</u> to rotate (become steeper as go up scale)." (3) Unlabeled diagram, with adjunct, indicating shift to right of saving as wealth increases.]

People get wealthy by saving. Have the habit of saving. Lower income people not necessarily in habit of saving.

Interaction between income effect and asset effect. In upper income classes the wealth depressing effect tends to disappear.

Level of income change: Not a very important effect, though does exist.

Interaction between level of income change and level of liquid assets.

Demographic:
[Two diagrams: one with age on horizontal axis and inverted U-shaped curve, and the comment, "Income moves through same cycle, with percent of income saved falling," and the second with family size on horizontal axis and negatively sloped curve, and the comment, "Opposite relation for contractual saving."]

A life cycle family may be constructed embodying typical demographic sequence. May be better way of combining these two variables with marital status.

Contractual saving starts at zero, not negative. Tends to be linear from origin.

Durable goods: whether married or not is what matters, not size of family. Liquid asset holdings on balance stimulate durable goods expenditure. Income change affects about like discretional [sic] saving.

Attitudinal:
If optimistic about general outlook, likely to save less than if pessimistic. Classify income change as "permanent" or "temporary," using people's expectations. Price expectations not happy about.

English data: 150–180 self-employed businessmen, farmers, etc. in sample. 1952–1953. May view decisions as depending on things that happen to consumer. [Equations centering on saving and durables (expenditures over £25), in relation to income after taxes, liquid asset holdings, mortgages, marital status, home ownership. Comments: Some results said "Not too meaningful." Also: "Some experiments using windfalls (pools, life insurance lump sums, etc.) and index of consumer assets owned by household: weak negative between durable expenditures and index, positive between durable expenditures and windfalls."]

How fit equations together? Relations between [forms of saving] and durable expenditures. Negative relation between durable expenditures and saving. Try [contractual saving or durable expenditures in durables saving] as free variables. Require simultaneous equations system, account [durables saving and durables] both endogenous.

Aggregation over individuals: [Age and employment] will change very slowly over years. Can be absorbed into constant terms.

Tarshis: Since are entrepreneurs, measure of investment yield expectations should be included.

RICHARD GOODWIN, "DYNAMIC ECONOMICS," APRIL 26, MAY 2, AND MAY 9, 1955

April 26, 1955:

Want to take up some of the ground covered by Mrs. Robinson in her forthcoming book. Treat the problems as I would treat them.

Long run macro-economics might be my title. Since Keynes' General Theory, this topic has dominated thinking. The short-run implications elaborated soon after the book appeared. The long-run part not worked out. Keynes' approach led naturally to recognition of sterility of nineteenth century economics re long-run developments. Harrod deserves credit for realizing what he was doing and for doing the right thing. Substituted a simpler theory as Keynes' was too complicated. First Harrod thought he had a cycle theory, and wrote a book, and then realized it wasn't, and wrote a book as a long-run theory. Return to Ricardian approach. Later claimed was both, but claim not substantiated.

A rich plum to be had. Are looking for the theory of capitalist evolution. Not easy. Not surely the result of hard work, the hard work required. I'm rather skeptical about possibility of expressing such a complex thing by a usable (simple) theory. This the historians' claim.

Mrs. Robinson has best seen the problem, and best equipped to deal with it. a/c long interested in Marx. Marx took over Ricardo's theory. Restated it and altered it, but didn't develop it much. Mrs. Robinson has one essay, "Generalization of General Theory," in her last book on Rate of Interest. [Equations relating saving to marginal propensity to save in relation to income; investment to accelerator in relation to income; condition of saving equal to investment; and "Harrod's proposition" that "Proportional rate of growth is a constant."] When Domar first thought of this, independently had no more than this, and wasn't long enough for an article. Had to pad it with material on monopoly, etc.

Very important, a/c different from Ricardian or any previous economic theory. Has continuous growth; does not lead to equilibrium. Harrod calls it equilibrium at all points, but not I. Is unstable, but not for Harrod's reason. Can't get off the path, if this is all the theory. Harrod mistakes fact that theory gives no answers when get off the path – for instability.

Is unstable, a/c curve moves off indefinitely.

Accords with experience of last 100–150 years for most of Western civilization.

But can't go on forever. Assumes no problem of land. Ricardo thought land would dominate.

Also ignores labor force. This I wish to consider.

Labor force adapts to this rate of growth or vice versa or both.

Also ignores technical progress. b [in Harrod's formula] is acceleration principle, which derives from given technique, reproducing itself on greater scale.

Harrod has tried to patch up to meet this. If technique given, roughly capital and labor, and output proportionate. [Diagram with income on vertical axis and time on horizontal. Series of positively inclined curves labeled "Labor" and, running through them, one such curve also labeled "income."] Productivity increases mean that labor curve drawn here rises more rapidly, but may not rise at same speed as Y [income] curve. [Smith wrote in brackets, "Labor curve must show full employment output that a rate of labor makes possible."]

Will population adapt to growth of output or vice versa?

Something like this must take place, if this is fruitful approach. Assume every unemployed person consumes fixed amount. Since theory above works through market for finished goods, connection with labor market must be through demand.

[Equations incorporating workforce, labor required, unemployment; with "given technique," "not explaining population," "taking population growth as given."] Will this do the trick? Theory unsatisfactory. Percentage rate of change of output not equal or proportional to growth rate of labor. Negatively related. [Amidst equations.]

Rewrite. A linear equation. [Amidst equations.] Exponential growth represented by straight lines. [Diagram with warranted income on vertical axis and income on horizontal axis, with several positively inclined lines, the first passing through point of origin, the others passing through horizontal axis to right of origin.] Unstable, a/c always tends to leave the equilibrium (zero point). Will never adapt to changing equilibrium given from outside. This a troublemaking mechanism.

[Mathematical expression] determines origin. Shifts curve to right at progressively accelerated pace. Tends to get further and further away from a shifting equilibrium point. [Smith wrote in brackets, "Doesn't this depend on whether the equilibrium point shift tends to change [warranted income] toward zero or not?"]

Mrs. Robinson probably doesn't agree with me that this is not adaptive.

What we need is a line negatively sloping, so will adapt toward equilibrium point. Stable system, with "pursuit curve," will pursue the equilibrium point.

[Diagram with warranted income on vertical axis and income on horizontal axis, with negatively sloped line extending from point of vertical axis above origin through point on horizontal axis to right of origin.]

[Equations spelling out stability conditions.] Don't defend it, but about as much statistical evidence as for others.

Now putting in labor supply means getting closer and closer to shifting equilibrium point. [Smith wrote in brackets, "This I don't see."]

This theory says economic adaptive mechanism limited to small changes, and main pattern set by biology or other forces.

May 2, 1955:
Look at problem other way round. Will labor force adapt to output.

[Equations embodying: wages proportional to output per head; simplified form of Malthusian theory (the Malthusian wage, however defined, is that at which population ceases to increase), income is exogenous.] Unstable system. But no feedback. [Income] not dependent.

Whole system stable about a trend given by the Harrod proportion. [Diagram with population on vertical axis and output on horizontal axis, with unlabeled positively inclined curve.] Get this result by finding solution to each part of the linear equation separately.

Still have to determine these values [in equation] to fit data at some point. This the general solution for includes all particular solutions, depending on N. Has one arbitrary constant. [Followed by equations and diagram.] If something else happens after it's started, as jump from a to b, will then again adapt (if population still sufficient to support this level of output, which we assumed independent of population.

Mrs. Robinsons's model (almost):

Technological progress. Two classes – capitalists and laborers. No natural resources. Homogeneous labor, single wage rate. Capitalist income what is left.

(Ricardo said land dominated everything. Not true in nineteenth century.)

[Equations incorporating determination of employment, wage bill, savings.] This the accumulation process, to which Marx devoted much attention, and others practically none.

[Equations representing saving as a function of marginal propensity to consume and warranted income – "Essence of Mrs. Robinson's version of Harrod's system" – and as function of technical progress, hence growth of productivity – "my own way" – with "My 'trick': Assume real wages keep pace perfectly."] Proportions of income to capital and labor remain same. Real wages rise. Both these effects observable in data.

Labor force? Adaptation assumes some kind of supply and demand of labor. [Equations, including average working class income as equal to pay divided by labor force, n.] Still a stable system, stable about a line determined by y, which is independent of n. However, as time approaches infinity, n becomes orthogonal to income, but employment is a decreasing proportion of output. Increasing unemployment. Dissipate higher wages among more people. Could have increasing leisure. Constant average standard of living. Not too satisfactory, but more correspondence to reality than most such models.

What if labor force and output independent?

Singer, H. W., "Mechanics of Economic Development," Indian Economic Review. [Equations.] Our concern is with y/n, standard of living. Or is it good to have y and n both growing at same rate? Unresolved problem of utilitarianism. [Equations regarding rate of growth of output per head.] But when are these values constants? Are they interrelated?

Singer gives figures which he says are kind of average of underdeveloped countries. [Numbers which give rate of growth of output per head to be –0.0005.] Substantially zero, and descriptively appealing.

Suppose seek minimum growth rate: Income must double in 35 years. [Numbers for variables associated with rate of growth of output per head thus required said to be "Very difficult to achieve" and "Not much you can do about it."]

Making not very capital-using investments doesn't seem very helpful. Need heavy capital-using investment for such country.

But: We've assumed saving proportional to income. But Kuznets short period data not proportional, while his long-period are.

Can argue that need not continue with savings proportional to income. Alter marginal propensity to consume, but not the average. Can't alter average, a/c standard of living already so low. But alter marginal, and gradually average will approximate the marginal.

May 9, 1955:

Singer's result depends on assumption saving proportional to income. But assumption very doubtful. Supported by Kuznets' long-period data, but his Great Depression data more like [diagram with saving on vertical axis and income on horizontal axis, with positively inclined line starting on vertical axis below origin and passing through horizontal axis to right of origin.]

Over a long period difference very great. In underdeveloped economy, observe simply low proportion of saving. Necessary that next stretch of curve have slope steeper than line to origin. If larger proportion of added income goes to saving, in long run will have adequate saving, a/c average will [gradually approximate] marginal.

Which way will people behave? Kuznets data cover ten year periods, and in such a period opportunity for depression to lower actual saving below level of what people try to save.

In a free economy, many think that rate of saving actually determined by series of accidents, not rational decisions. I agree.[2]

Certainly in a free economy people would not rationally choose to save constant proportion of income no matter what. Something else besides rational choice apparently in Kuznets' data.

[Through and amidst equations and diagram, shows rotation of savings curve; notes that "Underdeveloped country likely to need most of what is going for saving merely to equip new population with existing types of capital"; notes that long-run solution can be achieved by setting marginal rate of saving, and that "change of productivity has same effect on output as constant rate of growth of population."]

Mrs. Robinson the only one to realize how close Harrod's formulation is to Marxianism. But finding it harder than I would have suspected. The two-sector problem arises.

Robinson: Implicitly assuming capital permanently durable; gross and net profit identical. Capitalists save <u>all</u> of profit. Consume nothing unless earn by labor.

Note no economic regulator for saving.

No natural resources. No rents.

No technical advance.

[Equation.] Believe this would be Marxian version of Harrod theory.

Assume population subject to wage rate, for if population growth given, what connects the two? Population related to state of labor market.

Must distinguish employment from labor force.

Malthus stated population growth economically determined. Great contribution, but wrong. Population became separate subject from economics.[3] Demographers have come to a bad end. Predictions failing. They have ignored economics. Their equations conceal a lot of constants that are changing.

Is population growth partially determined by economics? If so, must allow for feedback into system, though would be simpler to take population growth as given. All right up to intermediate period. Take low growth rate of income compared with population growth: unemployment, population falls, saving increases, and have rise in rate of growth. Conversely also rate of growth adjusts.

Mrs. Robinson is trying to make this unstable system adjust to some kind of natural rate of growth. This the first problem.

The second problem: If income grows too fast (boom), population rises, savings fall, and income should fall. But in fact we get deficient effective demand. Above we assumed that was always sufficient effective demand to take off goods produced.

Robinson and Harrod say can conceive that rise of spending by capitalists balances fall by workers and vice versa.

This is only a growth theory. Shouldn't try to treat it like a cycle theory. Is not one. A cycle theory must be much more complicated.

[Equations involving explicit behavior of population, also labor force in relation to employment.] The higher the unemployment, the lower the wage.

So will eventually always grow at the rate appropriate to the growth of population. But for income actually to decrease wages must be greater than total output.

[Now considering normal unemployment.] Trend determined by labor force, and cycle about the trend. But cycle is explosive.

I haven't found a simple way to put down what Mrs. Robinson has in mind. Until I do, shall not be entirely happy about it. Yet is a very useful work.

Michael James Farrell, Problems in Econometrics, Downing Place, October 11, 1954–May 12, 1955

October 11, 1954:
    Don't read: Davis, Harold T. (two books)
                   Wold, Demand Analysis. Not suitable for beginners. Uneven quality.
                   Tintner, *** Not suitable for beginners.

    Suitable:
                   Good, but approach different from that of lectures:
                   Tinbergen, Econometrics
                   Klein, Textbook of Econometrics

                   Good, but not elementary?
                   Stone, Role of Measurement in Economics
                   Measurement of Consumers' Expenditures and Behavior in U.K. (1920–1938)

Don't wish to define Econometrics.
Movement started about 20 years ago to make Economics a science.

(1) Propositions not quantitative.
(2) Peculiar attitude toward laws.

"All swans are white" formulated as a law. But are black swans in Australia.
    Most scientists would formulate as "most," or "99%." Economists would say all white except when not.
    Typical proposition that all businessmen maximize profits, but need not be in the usual sense of profits.[4]

Scientists' method leaves you with measurable thing, profits, though not <u>all</u> businesses maximize. Economic theory formulated in terms of quantities not intrinsically measurable.

(3) Empirical information not satisfactorily incorporated.

Reform sought precise, quantitative expression of theory, embodying empirical information.

Required mathematical formulation of theory and use of statistical theory in applications. Result the opinion that mathematicians and statisticians were the proper practitioners. Believe too many of such refinements have come into the field.[5]

Believe <u>is</u> important reformulation of economics, and can be mastered, to large extent, without engaging in much formal mathematical and much statistical theory. I shall not give the statistical theory needed for the practicing econometrician (see Roy). The practicing economist can use econometric results without heavy dose of statistical theory. We shall deal with the basic econometric problems.

Winter term: Techniques for reformulating economic theory in suitable forms (programming, etc.).

This term: Building empirical information into economic laws.

Shall use scatter diagrams. Explained. Two variate case. Three variate case: Hours of sunshine, maximum temperature of day, and latitude. Three dimensions. More variates.

Curve fitting to scatter diagram. Common sense tells us would not want squiggly curve such as might go through all points. Straight line very convenient, but may be ruled out by theoretical considerations.

Consider $IR = k$: Electrical current x resistance = constant. [Diagram, unlabeled, with negatively inclined curve.] Try $I = k(1/R)$. Plotting $1/R$ gives straight line through origin [not given]. <u>Or</u> $\log R = \log k - \log I$. [Diagram with $\log I$ on vertical axis and $\log R$ on horizontal axis; negatively inclined straight line, unlabeled.]

Suppose: $y = a + bx + cx^2$. Transform: let $x^2 = z$, and have a plane in three variables.

Can usually reduce problem, by some dodge, to fitting of some kind of linear relationship.

How tell whether have the <u>best</u> straight line? We'll use lease squares, though isn't necessarily the best. To minimize [the sum of the squared deviations] has some advantages from view of statistical theory, but min[imizing] [the sum of the deviations] or something else might make as much sense.

$$y = a + bx$$

b is regular [negative?] coefficient

Want to be able to say whether relationship is closely observed, or loosely. Ask how much of variation in values of y is explained by regular [negative?] line and how much is left over.

$$\text{Variance: } (?d^2)/N$$

Residual variance: Measure deviations about regression line values.

Residual variance: proportion of variance unexplained by the regression equation. Conventionally have defined

$R = 1 -$ residual variance/variance

$R^2 = 1 -$ residual variance/variance = proportion of variance which has been explained

R^2 more useful than R, the corr[esponding] coefficient.

Next week: Sorts of empirical information we are likely to have, and how far it is likely to suffice for our purposes (prediction).

Sorts of data and problem of prediction.

Later: problems of demand analysis.

Then: problems of estimating consumption function; cost functions; testing hypotheses; simultaneous equation problem.

October 18: Missed lecture.

October 25: Aggregation Problems.

My article in current Review of Economics and Statistics is the precise treatment. Here give intuitive treatment.

Many aggregation problems, even in demand studies.

Aggregation over commodities: Answer: How must aggregate depends on what want to do with the result. If believe composite doesn't behave like single, treat separately.

Aggregation over prices: Time series of prices usually conceal variation from day to day and/or market to market. If D is straight line, average price lies on line. If D is curve, average lies on inside of curve. [Diagram labeled "D" with p on vertical and q on horizontal axis, with negatively inclined slope with three x's marked on it.]

Aggregation over individuals: We've been talking about market demand curve, or consumption function. Shall assume that know prices. Given prices and tastes, etc., seek relation between consumption and aggregate income.

[Diagram with C on vertical axis and Y on horizontal axis; nothing in space.] Hope $C = C(Y)$. Are justified in this hope? Economic theory assumes individuals behave rather consistent way. If $c_i = c_i(y_i)$, for individual can go to $C = C(Y)$, the aggregate demand function.

If individual income-demand functions differ, many problems.

First assume all identical: $c_i = c_i(y_i)$. [Can find total income and total consumption.]

Suppose have individuals at x's. [Diagram with C on vertical axis and Y on horizontal axis. Positively inclined curve, with two pairs of x's on it, bottom pair with arrow indicating upward movement, and upper pair with arrow indicating downward movement.] Keep aggregate income constant, and shift amount individuals. The C(Y) has changed, account top fellow decreased C more than low fellow increase

Linear demand function eliminates this difficulty, is said:

DeWolff, Economic Journal, 1941, Elasticity of Demand
Marschak, Review of Economics and Statistics, 1939, Personal and Collective
    Budget Functions.

But suppose not this: [diagram with positively inclined line commencing at point on vertical axis] but this [diagram with positively incline line starting at point on horizontal line.] Then original difficulty arises (Of course because function not linear through whole range – can't have negative demands.) As for "luxury." Inferior good raises same difficulty. Can't aggregate if have either "luxury" or inferior good. [Diagram, unlabeled, with negatively inclined line.]

Now suppose are differences in tastes. Distinguish systematic from random. Useful in exposition, though not always easy to apply.

"Individual" is the unit of decision – may be family or single person. Actually may be horrible mixture of decision makers in one family.[6]

"Individual" who is married man with six children will have different function than gay young bachelor. Is result of difference in tastes, and decision based on them. But we shall take as simply a matter of present tastes.

Country vs. city dwellers. Etc.

Such taste differences in some sense systematic, and circumstances usually quite easily measured.

Can think of classifying individuals appropriately and then constructing separate demand functions.

But if taste factor more readily represented by continuous variable, may do partial regression analysis, as for family size.

Result is to correct for systematic differences in taste, leaving only random differences. This set; $c = a_i + by$. [Diagram with C on vertical axis and Y on horizontal axis, with series of positively inclined lines intersecting one vertical line.] Only intercepts differ. This is the basic notion of subsequent analysis. Independently arrived at and used constructively by

Tobin, Survey of Demand Theory, Econometrica, October 1952
Malmquist, Statistical Demand for Liquor in Sweden, 1948

Assume that c_i = mean c and random deviation vertically. Assumed that could aggregate. But only true where have no people whose demand would be negative (if could be). But some individual demands may have negative range (if linear) even though mean function has positive intercept.

Suppose mean individual demand function (and each individual function) has intercept on horizontal axis. Can aggregate where all curves lie above axis, but not elsewhere. May result in curvilinear aggregate function. This is type of aggregate function that is relevant if consider data from budget studies. [Diagram, unlabeled, with numerous parallel positively inclined lines, with curvilinear function drawn through them.]

$$C = C(Y).$$

Call it "budget" function. Differs somewhat from genuine aggregate function: $C = C(Y)$.

Practical implication of budget function: Has been assumed that gave guide to individual demand function shape. Not so, as seen above.

Once have budget function, can think of population as having identical demand functions, just like the budget function.

Assumption that random distribution of tastes independent of income has been used here.

Distribution of tastes difficulty important only if wish to go between aggregate or budget demand and individual demand curves.

Rationing: Suppose know individual demand without rationing. But not all reach ceiling at same income level. Results in [kinked] budget function. [Diagram with C on vertical axis and Y on horizontal axis, with kinked budget function.]

Substitute commodity: Budget [function] again curved. [Diagram with curved budget function.]

Next week: (1) method of deriving a priori shape of utility functions for individual comm[odity]; (2) how deal with durable commodities in demand analysis.

Two following lectures: consumption function.

November 1: Missed. Aggregation Problems dealt with.

November 8:
Consumption function. Important for policy – predicting level of employment. Why notorious failures in U.S. re postwar employment? Klein, "Postmortem on Transition Predictions of U.S. National Product," Q. J. E. ? 1947 ?
   Appear to get good fit of consumption against aggregate income.
   Snags:

(1) Identification problem. Both consumption and investment depend upon level of income? How do know are not getting investment function? Not easy to tell.
(2) Consumption of durables. Changes in income cause changes in consumption, but if durable good, especially may rise by greater percentage than income (stocking up). War draws down stocks. Stability does not extend to durables, which depend also on stocks, or previous levels of income, or previous levels of expenditures on durables.
(3) Liquid assets
(4) Price. (a) But idea of consumption function is that relative prices of spending and saving don't matter much. Likely is so. See Harrod, Towards a Dynamic Economics. In any case, not too clear what is direction of interest effect. (b) Is consumption a homogeneous commodity? Consumption is least homogeneous of all commodities. Saving might be closer substitute for some commodities than other commodities are.
(5) Serial correlation between predetermined variables. More important worry than serial correlation between errors, I believe. May get high correlation between income and consumption, but low between changes in income and changes in consumption. First differences reduce both total variance and variance explained, but not the unexplained variance unless is correlation of errors. The apparent correlation between income and consumption is open to some doubt, as may be due to fact that each variable has high serial correlation.

Other nests of problems.

(6) Long-run savings ratio. Kuznets found savings a constant proportion of income. Cyclical data gives smaller slope that would have positive intercept. [Diagram illustrating this.]

Hicks (Trade Cycle) suggests differences purely due to lags. Consumption rises and falls less, in any one year, than over longer period. Alternative: Modigliani, "Fluctuations in Saving-Income Ratio," Proceedings (Studies?) Income and Wealth (1949), and Duesenberry (book) suggest that consumption depends also on previous highest income. Modigliani: $C = a_1 + a_2Y + a_3Y_n$. Duesenberry: $C = a_1Y + a_2(Y_2/Y_m)$. Kind of ratchet effect. The short period consumption function rises a/c of habit formation.

I applied similar method to consumption of habit forming commodities, . . . Interesting, though not conclusive results.

Both lag and irreversible explanations seem adequate, but data inadequate to test hypotheses.

(7) Budget study data: Averages for income groups. [Diagram with consumption function for whites above consumption function for negroes.]

Get different results in different communities. Although marginal propensity to consume apparently same, whites consumed more at given level of income. Also found between cities, so couldn't explain on racial psychology. Then suggested that affected by spending level of neighbors. Very attractive hypothesis, to theorist as well as re the data. In one form, made consumption depend on one's position in income scale. Feel this less attractive than consumption dependent on own income plus average level of expenditures in community. But can't distinguish between the two from data. See Duesenberry.

Tobin then pointed out that liquid assets provided sufficient explanation for these hypotheses: $C = a + bY + cL$ works just as well. Tobin-Duesenberry controversy. Inconclusive results.

Next time: New (and promising) theories.

November 15: Consumption Function
New work, providing unified solution for whole set of problems. Began with Roy Harrod's theory of hump saving, in Towards a Dynamic Economics. One saves in order to dissave when old. If were only reason, no net saving in economy, if people clever about it.

Why is there?

(1) Not only reason. Save to leave to children.
(2) Save to take beyond expected retirement period, a/c of chance of living longer. People, will, on average, leave a certain amount of money in their hump, even though don't intend so.
(3) Population increasing. More young savers than old dissavers.
(4) Income per head increasing. Dissavers based on lower level of income than savers.

Reasons (1) and (2) more or less offset by plans of heirs to dissave their inheritance over the period of lifespan. Points (3) and (4) have received most attention in new consumption theories.

People who've developed the new theories: Milton Friedman and Margaret Reid. Modigliani and Blumberg. Developed independently.

My version here is a composite and a selection from these two.

Set up model in which all individuals are alike. Assume no time preference. Harrod shows important distinction between time preference and diminishing marginal utility of income. Latter would lead to distribution of income evenly over time. Not excluded by assuming no time preference.

No rate of interest enters theory.

Chap will work N years, earn Y^e <u>average</u> income, and live L years from time starts working.

$$C = N/Y \ (\overline{Y}^e)$$

Individual consumption function has slope N/L. 40 years work with 50 years L gives slope of 0.8.

What of long-period savings-consumption ratio? Effects of given rate of increase in population and of certain rate of increase of income per head turn out to be the same. So deal simply with rate of increase of income. Get 12% saving rates from 3% income increase. Very satisfactory.

Aggregate long period function has slope of about 0.9 rather than 0.8. [Diagram illustrating both slopes.] Depends on rate of increase of income, note.

What of cross-section data? Each individual plans to be on individual line. Would find him there if plotted consumption against $\overline{Y}^e$, rather than against Y.

$$Y = \overline{Y}^e + n.$$

(1) Expect year to year fluctuations in income. One bad year won't affect expectations over lifetime and thus won't alter consumption.
(2) Unexpected deviation – windfall gain or loss.

Expected income becomes $N\overline{Y}^e + n$, and spread all of this over life span. Increase current consumption by N/L, where L is of order of 50.

Deviations affect income more than consumption. People find selves on more sloping line [illustrated on diagram].

Study of slopes of budget study functions by occupations. Where n small, slope of budget study function is large; where n large, slope is slight (notably farmers).

Different consumption functions in different communities: Suppose assume deviation from individual produced by n.

Community with lower $\overline{Y}^e$ has lower line. n becomes variation from average of community, setting up two lines of smaller slope. Depends on assumptions that Y greater than average associated with positive n.

I'm not so happy about this argument. Plausible, but no more. A third hypothesis for this case, in itself not evidently better than other two. But unifies with other cases.

Short period marginal propensity to consume:

Aggregate change in income

Aggregate change in consumption

(a) Individual change in income may be expected, with no effect on dC
(b) Windfall dY. $dC = dY/L_{-t'}$, where L_{-t} denotes <u>remaining</u> life span. For whole community, L_{-t} may range from 50–20, say 30. Marginal propensity to consume is about 0.03.
(c) Unexpected, and causes revision in expectation of income.

Raises $\overline{Y}^e$.

If $d\overline{Y}^e = dY$

$dC=(N_{-t})dy/L_{-t}$. Assume $20(dY)/30 = 2/3$ for marginal propensity to consume. Has been shown that over cycle is likely to be about 2/3.

Quite possible that $d\overline{Y}^e > dY$. May get marginal propensity to consume > 2/3, or even > 1. A really iconoclastic result.

Raises hob with use of marginal propensity to consume for prediction, as makes apparent that expectations can cause it to vary without limit.

Summary: [Diagram with three consumption functions: long-run, individual (true) slope about 0.8, and budget study.] This true individual function only remotely related to any of the things we measure econometrically. Slope of budget study function depends on correlation between actual income and expected income. Differences between budget study function slopes measure differences in this correlation.

Cyclical consumption function looks like budget function, but slope depends on elasticity of expectations.

None of previous approaches to consumption function have measured real consumption function. Have measured three other things. Our forecasting attempts have been pretty silly, naturally.

None of this work has yet been published.

Two lectures left this term

(1) Statistical measurement of supply, cost, production functions, etc.
(2) Simultaneous equations, verification, prediction, etc.

<u>November 22</u>: Estimating statistical cost and supply functions
Much less satisfactory than estimates of demand and consumption functions. I'm not completely nihilistic.[7] Have been some recent attempts, rather extensive,

in terms of a new set-up. Involves rewriting of economic theory to make it suitable. One of main tasks of econometrics.

Today will survey things that people think can do, which find we can't.

Cost functions:

(1) Analysis of what costs to run particular piece of machinery in particular way is not really an economist's problem. Results useful, but often trade secrets.

(2) Short period cost function of a plant. Observe plant over period of time, noting cost and output. Plot one against other. Which costs do you plot? How treat depreciation?

If very short periods of output fluctuation, plant and labor force not well adapted to output. Can't have too short a period. If longer period, less observations. Technical progress occurs over time.

If should get this function, doubt its interest is great.

(3) Long period: Comparing costs of different sizes is what it comes to, since not an historical cost concept. Technology varies over time. Rostas, Productivity, Prices and Distribution in Selected British Industries, has done this sort of thing. Have to use Census of Production data. Drawbacks:

(a) data always presented in grouped form (at least three firms)
(b) small sample – number of firms much smaller than number of consumers
(c) efficiency of management as important as scale
(d) age of firm important to costs, a/c (i) age associated with technology and (ii) older firms likely to be spending less on depreciation
(e) depreciation charges hard to handle.

Supply prices: Considerable practical interest. Data for total production and price easily obtained, for industry.

Industrial product: Observations will depend on pricing rule followed. If follows marginal cost equal to price rule, a la Joan Robinson, would give supply curve of theory. P. W. S. Andrews, Manufacturing Business, and articles, Hall and Hitch paper not representative. Andrews differs from their conclusions.

Andrews: no change of selling price in response to change in demand in short period. Let orders pile up. Supply curve in a sense perfectly elastic, to capacity. [Diagram with parametric price line, with two negatively inclined lines intersecting it, and point "a" on price line midway between points of intersection.] But if demand curve too far to right, observation at "a" – capacity output at fixed price.

But he also argues that they do change their prices, to pass on to consumers changes in function costs. [Diagram with one negatively inclined line intersecting four positively inclined lines drawn through scatter.] Scatter generated by function cost shifts, not the demand shifts of classical theory.

Agriculture: Different story. Farmer behaves more like perfect competitor. But another problem: A large random variation a/c weather makes measure of demand easy. Large random shifts of supply curve. [Diagram with one negatively inclined line intersecting three positively inclined lines, points of intersection marked with x's.] But supply curve hard to catch from time series data. Can abstract from weather?

Suppose consider quantity of resources devoted to a crop. Related to expected output, which relates to expected price. A definite step forward to relate acreage to price. Extremely difficult to measure non-land inputs, a/c many devoted to a variety of crops and not separable.

Experiment at London University by Morton and others: Used international cross section data for estimate of production function. Plot input-output data as though were budget function.

But: Census of Production data. Probably a high degree of aggregation, to get comparable industries. Not sure how types of product in each country's industry are weighted.

Small number of observations (countries with suitable data).

A priori notions of shape of production function seem pretty shaky.

How measure input?

Labor easiest, but: man-days, years, or hours? Output per man-hour may go up when output per man-week goes down, a/c length of week.

Skill? Industriousness of labor? British workmen notoriously easy going, compared with either of two continents.

Capital inputs: How measure? Morton et al. reduced to using number of horsepower installed. Is this even a reasonable approximation to manufacturing industry? Most will have power operating machinery. Differences in capitalization not likely to be reflected in degree of horsepower installed, I believe.

Then what becomes of your production function? No good if can't measure other inputs than labor.

Why do we need these estimates of cost functions? Welfare economics – long period marginal cost needed. Close down old mines? Build new power station? Can be done, but is job of engineer, not econometrician. Intensive study of costs of typical added firm may be of use in some industries.

Next term will talk of other attacks on the cost and supply functions.

References: Johnston, Oxford Economic Papers, 1952 (not in spirit of intense recommendation); Chenery, Econometrica, 1952.

Next week: problems of prediction, verification of hypotheses, etc.

November 29:

Simultaneous equations: Enormous importance attached to this in U.S., especially by Cowles Commission. Agree that the problems are important.

Have you a demand curve? [Diagram with price and quantity on axes and negatively inclined set of four small circle-points.] A line through these points implies a series of shifting supply curves. [Diagram with negatively inclined supply curve intersected from above by three negatively inclined – but more inelastic – demand curves.] But suppose were this? Suppose: [Diagram with negatively inclined demand curve residing between two parallel lines of circle-points and intersected by one positively inclined supply curve.] Perhaps demand shifts small and supply shifts large. Then could estimate demand pretty well. If vice versa, estimate supply. But might get big cluster that is mixture of both.

This the identification problem. Need to be pretty sure what relation data do represent.

Shouldn't worry any sensible econometrician, a/c has already taken care of this sort of thing.

Bias:

Suppose know that supply pretty constant and demand shifting. When demand shifts a lot, supply shifts a little a/c common factors, suppose correlated disturbances. Then slope of regression will be overestimated. Bias which may arise out of simultaneous satisfaction of relationships. Do have to worry about it. But sensible econometrician would worry about it before fitting his curve.

Another kind of bias: [Diagram with positively inclined line residing between two parallel lines of circle-points.] Suppose fit regression minimizing sums of squared deviations taken vertically .... But suppose quantity bought depends on price, with random element in quantity. Then should minimize squared deviations horizontally. If deviations small, two relationships will not differ much. Very important to know which variable regard as dependent and which predetermined.

Applies in two ways: First way mentioned above. Perhaps an overstatement, a/c traditional least squares people had thought of this problem before simultaneous equations people brought it up.

Second way: Large var[iations] in demand, small in supply. No correlation between shifts. [Diagram with three weakly drawn supply curves intersected by six demand curves. Intersections on middle supply curve marked with x's; others with circle-points.] However, if demand not perfectly inelastic, true points would have been the x's. Sums of squared deviations ought to be minimized in direction of <u>demand</u> curve, else bias. Liable to get it whenever have two simultaneous relations between the <u>current</u> price and quantity variables.

Sensible econometrician may have taken all the obvious precautions, and still be subject to danger of bias if use least squares. But by no means clear that this sort of bias is important.

Burdens of simultaneous equations: (1) Increased computational effort. (2) Increased variance in case of limited in form technique.

Variance vs. bias: [Diagram with two differently-peaked distributions.] One method of choosing is to take smallest mean square deviation between each value and true value. Other methods.

Hildreth, M. R. Fisher, and R. Bergstrom have been investigating bias vs. variance problem.

Doubt that extra trouble usually worth while. If want to follow up simultaneous equations method:

Klein, <u>Textbook of Econometrics</u>

Cowles Commission Monographs number 10? and 14? (more readable)

Serial correlation of errors: Quite likely in time series. Doesn't bias estimate of parameters, but reduces estimates of standard errors of parameters. May lead to overconfidence. Durban and Watson, <u>Biometrika</u>, 1951? No very good answer found. If form of serial correlation could be postulated, could improve estimates of standard errors. But if make mistake in postulation, could through [sic: throw] <u>way</u> off, Watson concluded.

I wouldn't worry too much about it. All statistical analysis assumes that are some systematic factors and some random (caused, of course, but by multitude of factors). If can approximate experimental sciences, can use their refined techniques.

But in econometrics, error terms large and not caused by multitudinous causes, but simply by causes not measurable.

We must restrict ourselves to methods that do not depend on precise knowledge of form of errors. Argument applies also to simultaneous equation methods.

Time after time regress equations with high correlation with time series data have given poor forecasts. Unwise to rely on standard errors of forecasts. Always the chance that some dormant variable has suddenly sprung to life.

Dormant: constant, or highly correlated with another variable. If wakes up later, poor results.

Idea that: Tests of econometric results should be success in predicting. Correlation not the important thing.

Professional statisticians immediately annoyed. If two lots of data originally used, would have shown up the difference. [Diagram with consumption and income on axes and with two groups of circle-points: one with positively inclined line of central tendency drawn through them; the other off to the side.] But suppose: [Diagram with two groups of circle-points: with (1) two positively inclined lines of central tendency and (2) one horizontal line of central tendency.] Theoretical statistics approach might have given quite respectable correlation coefficients. But find systematic residuals as warning of activity of dormant variable. Still, we're humanly likely to overlook systematic pattern in residuals. Prediction method is check against these. If prediction the test, may be less fussy about using some of more refined measures. Prediction as the touchstone enables us to dispense with a lot of complicated mathematics.

Some may have overreached selves. Predict plausible results. Cf. with "naive" model. Fit regress, extrapolate, and miss a bit. Take as naive model that consumption is same as year before. [Diagram with consumption and income on the axes, with positively inclined line through circle-points and with circle-points for 1946, 1947, 1948 off to right with 1947 to right of 1946 and 1948 above 1947.] In 1947 superior to econometric model. But fallacious a/c econometric model not given any of the up-to-date information that is available to naive model.

Next term a non-stochastic part of econometric work. Rewriting of economic theory in useful form: programming, activity analysis, etc.

January 17, 1955:
This term: Reformulation of economic theory, primarily to make it more tractable from empirical point of view, to make possible useful inferences from kind of data likely to get. Also indicates some of shortcomings of existing theory.

Is a reformulation, not a revolution.

Books: Subject still in pioneering state. The most important:

Koopmans, ed., Activity Analysis of Production and Allocation

R. Dorfman, Application of Linear Programming to Theory of the Firm

Charnes, Cooper and Henderson, Introduction to Linear Programming

Part I (Cooper and Henderson): Intended as non-mathematical introduction to Part II

Part II Quite mathematical, but good. Some have to read this in order to understand Part I.

This sort of reformulation has been developed by several somewhat independently. Have, not a theory, but a set of special problems, more or less related.

Names: Activity analysis, linear or mathematical programming. Dorfman, Charnes et al. have essentially the view of obtaining mathematical solutions (numerical).

Input-output analyses and matrix multiplier are also somewhat related.

One common feature: deal with many variables.

Little has been done in traditional theory to analyze dynamic behavior of firms. Assumptions that seek to maximize either short period or long period profit are poor.

When consider many periods, need many variables. Much harder than static equilibrium analysis. Thus tendency for theory to be static.

The multi-product firm, or multi-factor analysis, has also been neglected. International trade theory limited to manageable number of variables. Keynesian also.

Some theorists have many variable solutions in orthodox way:

Tintner, mostly Econometrica during war.

Functions hardly can be specified or empirically applied. Formal mathematics very general.

Trouble is that although the existing theories would produce good results if we were all very clever and had enough and proper data, we are short on both counts. Have to cut something out, in order to make thing tractable. What we do is, generally assume that functions are linear.[8]

Thus enables to handle many variables. Failure to handle many can sometimes lead to serious error.

Justification:

(1) The great limitation of the marginal analysis is that its equilibrium conditions give only local maximum. As long as stick to local maximum and small variations, linear functions make pretty good approximations.
(2) Needn't assume straight line over whole range. Need only be piecewise linear. Can approximate any function you like by making pieces small enough. At expense of added labor cost.

    Accuracy of measurement so limited that small number of linear segments likely to be plenty, in empirical work.
(3) Illustration: Stranded in small village, and short of cash. Only one hotel, and can't afford to both spend the night and have dinner. Must choose. But can't divide bed and dinner, as marginalist analysis assumes. [Diagram with dinner and bed on axes, with budget line and indifference curves assumed linear; with conclusion "choose bed" – corner solution – because on highest indifference curve.] In fact, most of examples of consumer choice are of this type.

Are far more commodities that you don't consume than there are that you do. The corner optima are important.

In really good theory want <u>both</u> corner maximum and tangential. But shouldn't ignore corner maximum.

Could have piecewise linear indifference curves. One kink gives three possible optima (aside from infinity if tangency occurs). [Illustrated on diagram.] The two hotel case.

The kink not uncommon. Have to choose between one car and two. Many commodities indivisible.

Most applications of this theory to production. Not clear that linearity assumption a dead loss.

(1) Multi-product and -factor theories of firm
(2) Multi-period theory (a real advance)
(3) Berlin air lift problem (don't know whether did use this technique, or whether simply thought would have been good idea to do so). Reduce present food to fly more airfield material to give more food in future?
(4) Transportation problem. Shipping. Coal deliveries.
(5) Gasoline blending. (Sought to find out what oil company should be told to do, if it were interested in knowing.)
(6?) Input-output.
(?) Goodwin's matrix multiplier.
(?) Inter-country and -regional multipliers.

<u>January 24</u>: Formal set-up of activity analysis
Economy: Perfectly general – national, regional, world, firm, part of organization, etc. Set of processes making some things into other things.

Commodities: General factors also included. Economy changes commodity into other commodity. i from 1 to N.

Activity: Changes certain commodity into others. Constant returns to scale. Inputs and outputs proportional.

Output of ith commodity from jth activity is $a_{ij}x_j$.
x_j = level of jth activity . . .

If $a_{ij} = 0$, ith commodity not consumed
If $a_{ij} > 0$, ith commodity output
If $a_{ij} < 0$, ith commodity is consumed, or used as input

[N activities plus technology matrix. Output equation tells what can do with the economy, from technological considerations.]

Commodities:
Primary – inputs to economy as a whole.
Intermediate – produced and consumed again.
Final – produced, on balance.

Can classify commodities, from technology matrix.

Problem:

[Suppose some of primary commodity in perfect inelastic supply. Subject to this restriction, seek to maximize net profit Linear function of x's, to maximize subject to set of linear inequalities on x's, and condition . . . Could generalize still further by having prices on scarce functions . . .]

Back to other problem: Assume perfect markets. Don't know until have solution whether commodity is primary or final.

In many problems, the prices may refer to imputed prices which the decision making authority lays down. Prices existing in the market may not be the ones that wish to consider.

Generally, certain commodities will be incapable of sale, purchase, or either. Generally, intermediate will be neither capable of sale or purchase . . .

Can't be bought . . .
Can't be sold . . .
If can be thrown away without cost . . .
If must pay to dispose of . . .
Neither bought nor sold . . .

Orthodox activity analysis approach: Koopmans assumes can classify in advance. Must provide disposal activity for each intermediate activity . . . Amounts to what we've been saying.

The equalities can be used to eliminate the unneeded disposal activities.

Variant: All primary commod[ities] scarce, and want to maximize prod[uction] of final commodity.

Reverse: Suppose production targets exist, and wish to minimize cost. Maximizing . . . will minimize cost, . . .

More general: Certain final products required, sale of certain by-products permitted, revenue therefrom offset against input costs. May be minimum cost of even a positive profit.

May be restrictions on both inputs and outputs. Have to take care that inequalities are consistent, or may be asking the impossible.

Difference between activity analysis and linear programming. Activity analysis gives this conceptual setup. When reduced to the mathematical form, is same as linear programming. Some problems that can be reduced to this form can't be fitted into activity analysis framework. Linear programming contribution is the arithmetical solution. (I will postpone consideration of solution procedures, and may not actually get to them. Liable to be dreary in classroom.)

Realism of assumptions:

Assume finite number of activities. Not continuous substitution. Seems rather a good thing. More realistic, usually.

Approximate each other. Continuous curve is the approximation to reality. [Diagram with land on vertical axis and labor on horizontal axis, with combination of negatively inclined curve and function comprises of straight-line segments.] But how do get the straight lines joining the activity points?

Assumed: (1) Can allow x_j to vary continuously.
(2) Add outputs of two processes.

From these two assumptions it follows that any point on line joining two activities is weighted mean of end points.

Is this type of substitution possibility an improvement on traditional theory?

January 31:

Last time: activity analysis assumes must choose between finite number of processes, so continuous substitutability not available. Seems a point in favor of activity analysis – like reality.

But look more closely. Wish to cf. assumptions and results of activity analysis and orthodox theory. Easier to make points with diagrams.

[Diagram with land, y_2, and labor, y_1, on axes, and negatively inclined function comprised of straight-line segments, and with line from origin to point beyond line.]

Have output target – real goods, or sum of money.

Types of farming give different points.

r   = unit level revenue
p   = fraction of unit level
pr   = total revenue
py_1 = input labor
py_2 = input land

pr	py_1	py_2
$(1-p)r$	$(1-p)y_1$	$(1-p)y_{2'}$

r

Weighted mean is input of labor.

Any point on straight line may be had by varying price.

Since wish to minimize use of inputs, always prefer point or mixture on line nearest axes. Any point to northeast of line joining two other points is inefficient.

Looks much like marginal analysis. There is a kind of continuous substitution. Land and labor can be used in any proportions we wish.

Is the divisibility assumption that is responsible for this result. (Also additivity used.) This the one that detracts most from the realism and usefulness of activity analysis.

How valid is additivity assumption? May sometimes believe the curved isoquant more realistic than the straight line. Manure from dairying may help wheat farming.

But can always approximate this curve by adding line segments. Introduce "fertilized wheat farming" which consumes the intermediate commodity manure. Additivity assumption thus not seriously restrictive. Better specification of technology will give the degree of realism that is wanted.

Let level to which process is carried be dictated by amount of machinery. Only points on northeast boundary are efficient.

Manufacturing firm. Want to consider inputs of labor and output of commodity. [Diagram of first quadrant, with product on vertical axis and labor on horizontal axis. Straight line from origin to northwest labeled "I f machinery unlimited", with some points marked off line to west.]

We want to represent increasing and decreasing returns. This has only decreasing.

Diminishing returns a/c limitation on machinery. If unlimited machinery would use only first profess (constant returns). Accords with real life, in short period cases.

Long-period diminishing returns a/c either: (1) some factors not homogeneous (extra ones less skilful or convenient). Really have a set of different factors, each with [diagram with kinked function] supply curve.

Or: (2) In a sense, a rising supply curve, with factor measured in efficiency units. Same wage per man, but higher wage per efficiency unit. Fits well into marginal analysis, but not into activity analysis.

The first method of representing factor scarcities is more realistic anyway.

Increasing returns: Why not in our model? Divisibility rules it out. Prevents us from representing increasing returns. [Unlabeled diagram similar to last diagram but with kink.]

Dorfman reformulates theory of firm in terms of activity analysis. Isoquant from various processes. Limited amounts of land and labor. Maximize net revenue along vector drawn. [Diagram with land and labor on axes, isoquant and vector.]

Criticism: nothing to allay our fears re additivity. No possibility of intermediate products affecting in any way. Has restricted self to a very simple model.

His second model assumes linear supply or demand curves for factors or products. [Seek different maximum.]

Requires addition of quadratic form to x's. Quadratic programming, not linear. Could use such device to represent increasing returns? . . .

Could thus represent indivisibilities of inputs?

If concerned with indivisibility that occurs only once, then a good answer. [Diagram with declining average cost curve and decline marginal cost curve.] But if more than once: [diagram with sequence of three declining average cost curves and corresponding declining marginal cost curves.]

Model becomes complicated a/c each supply curve applies for different levels of y.

Summary: Additivity not too restrictive. Divisibility is. Likely to be important part of problem affecting firm. The linear supply function doesn't seem to help much.

Alternative: Let yi take only integral values 1, 2, 3, 4. Can always work out numerical solution by comparing results of alternatives available.

But another thing to produce an illuminating solution, not simply a particular solution to a particular problem. Doesn't indicate what sort of answers will get from a fair class of problems.

Shall argue later that programming has more to offer in dynamic cases than in static, so more to be gained there, even if this difficulty still holds.

February 7:

Should have distinguished more clearly between two sorts of use of activity analysis: (1) to solve a practical problem; (2) to be illuminating in way in which economic theory is. Did not want to be disparaging at all re first use.

Theoretically illuminating means? A kind of prototype solution – type of large class of problems. Seeks elegance and illumination. From this point of view perhaps too much has been claimed. Possibility in complicated international trade problems may help.

This time: Simple dynamic activity analysis problems.

Cahn [sic?] warehouse problem: Assume costs of use can be regarded as overhead. Buy and sell wheat at same price. No handling costs.

		Buying $x_i(t)$	Selling $x_2(t)$
Final commodity:	Net sales of wheat y(t)	−1	+1
Intermediate:	Stock of wheat z(t)	+1	−1

Activities: Buy. Sell.

Prices: p(t)

Maximize . . .

Constraints: [elaborate set of equations]

May introduce disposal activity x_3 which leaves warehouse unused.

This is basic framework of a fundamental economic problem. Can elaborate. When put in buying and selling costs, turns out that either buying or selling must be zero [Smith added: "(at given time?)"]

A step forward from static theory.

Berlin airlift model: Cowles Commission Monograph, but was it related to the actual solution of the real problem? In any case, was a simplified version. My version simplified further.

In a given time period

[Matrix with alternatives: (1) Fly in materials to enlarge field. (2) Fly in food. (3) Disposal. Results: (1) Final Commodity: Deliveries of food. (2) Intermediate commodity: capacity of airfield, at beginning and at end or period.]

Usually good idea to write out matrix for only one time period.
Use of units that gives 1's clarifies thought.
[Elaborate set of equations re maximizing subject to constraints.]

Linear programmer apt to say that result obvious, and needn't use all this elaborate procedure. Some people feel the conceptual apparatus of intermediate commodities helpful.

Can sometimes use intermediate commodity to eliminate disposal activity. Can here eliminate $x_2(t)$ still better.

[Elaborate set of equations, including notation that Smith believed at least part of it to be incorrect, and statement, "Obvious that will want all $x_3 = 0$, as have negative coefficients."]

Consider third term. As coefficient of $x_1(t)$ is greater or less than zero, will seek x_1 as large as possible (1) or as small as possible (0).

Appears unrealistic, a/c all-or-nothing. Depends upon assumption that profit function linear in y's: attach a constant p to deliveries of food, whether large or small.

Cf. rate of interest: Analogous to case in which have genuine time preference, but no diminishing marginal utility of money. See Harrod's article in Towards a Dynamic Economics. Absence of time preference means same schedule of marginal utility for money each year.

In airlift problem we've put in time preference and left out diminishing marginal utility, yet the latter the more important factor here, as in int[erest] theory. Time preference alone will leave you building air fields now, with no food until later.

With p a linear f(y), back to quadratic programming. Could set up problem with minimal quantities of food at higher prices than rest of food in any given year. By generalization could keep whole problem linear, yet introduce approx[imate] to whole downward sloping demand curve.

Further generalization: Might consider possibility of: (a) having stocks in Berlin to begin with; and (b) putting food into store.

Next week: How we can improve theory of firm.

February 14:

Using activity analysis for multi-period analysis of firm theory. Article in Econometrica last year. Will talk here in more intuitive way.

Problem simple: Firm theories, early, made profit $= R(x) - C(x)$. Maximize re x. But question arose whether relate to short period or long. Answer in either case, but don't believe businessmen maximize either short or long period profits. Latter would involve setting up business on much larger scale than is usual, and run at loss for many years.

Businesses do maximize some weighted sum: . . .

Another feature of business conditions as in Andrews's Manufacturing Business: existence of a market attached to the firm, but not Chamberlin's case. He assumed would lose customers gradually as price raised, but not all, a/c their irrational preferences. But majority of businessmen don't sell consumer goods to consumers; they sell components to other businessmen, or consumer goods not highly differentiated. Businessmen can't afford, in general, irrational preferences, for indefinite period. But is surprising how firmly businessmen think of having market attached to them. Is essentially a matter of having customers who might do as well with one's rivals, but don't want the trouble of changing over. In real world, may regard products as homogeneous, but customers are attached by inertia resulting from intangible costs associated with change.

If firm gets out of line, will lose a proportion of its customers, those with smaller changeover costs or quicker reaction. Number you lose depends on how long price disparity persists. "Goodwill."

Fits rather neatly as an intermediate commodity. Must have "goodwill," built up and for use. Analogous to airfield capacity. Determines scale of operations.

Difference in reaction to upward and downward price movements: may be easier to drive away than to lure back. Can easily fit in this kink. May also easily fit in difference between short-period and long-period elasticities of demand. May get quicker response to price rise and slow one to fall.

Technology matrix. (Have used implicit sort of cost function – linear, and output can't be stored.)

[Matrix formed by Nature: (1) Selling, (2) Raising, (3) Lowering, and (4) Ignoring, horizontally, against (1) Intermediate Market, (2) Final, (3) Primary Costs, and (4) Intermediate Market, vertically.]

Have no control over nature. Includes: Exogenous changes in goodwill, competitive price (leaves size of market unchanged). In practice manufacturers ignore demand a good deal. Our model makes it rare, but a/c uses linear cost function. In real life marginal costs rise rapidly a/c limited capacity of plant.

[Given cost, maximize; have quadratic term.]

By holding nature constant, get nice conclusions: [Under certain conditions, businessman follows policy of matching competitive price. Have removed short-period elasticities. Find the gross profit margin, quasi-long period elasticity, percentage loss of market from one period rise of price.]

Whether take quick profits or not turns on magnitude of variables in this expression.

Hicks says that if short-period elasticity is less than infinite, and if interested somewhat in short period, would raise price somewhat in present. (Oxford Economic Papers last year.) But not quite valid.

[Notes that price increase is a function of elasticity but also of discount rate to take care of both time preference and uncertainty.]

One week skipped.

February 28:

Input-output analysis: Special case of programming. Commodities: $y_1 \ldots y_n$, y_{n+1}: Ax

Peculiarities of technology matrix: Only y_{n+1} is to be primary. Permitted only n activities.

For each of n activities, only one commodity for which output positive. Only one possible way of producing a given commodity.

Given $y_1 \ldots y_n$, choose x so as to maximize y_{n+1} (minimum absolute value of y_{n+1}). Of course, only one set of x's possible.

[Mathematical notations omitted.]

Might seem useless exercise. Interesting part of problem left out. Much justice.

But argued that may reduce problem of optimal production of n commodities into two sections, one considering all possible methods of production. Having maximized, are left with second step of main problem. Suppose the businessmen have done maximizing, and I need only find out what levels of activities are required to produce given bill of goods.

Theorem that if retain general set-up of problem, and assumption that any activity produces only one positive output, then allowing any number of activities, will conclude with only one activity producing each output.

Let $n = 2$.

[Diagram with four quadrants, with y_2 on vertical axis and y_1 on horizontal axis. Groups of circle-points near origin in first and third quadrants, with lines through second quadrant connected two pairs of points, one pair with x's rather than circle-points; and with ray emanating from origin in second quadrant to darkened circle-point labeled "objective."]

Activities produce y_1 and consume y_2 and y_3 or produce y_2 and consume y_1 and y_3. Run each activity at such a level that y_3 input is constant (at unity). Seek minimum input of y_3 for given output.

Seek set of activities that will get as far along ray to objective as can be. Will be one such line. And will be same line, whatever ratio of y_1 to y_2 we choose. The same two activities will be optimal. (There are no activities producing both y_1 and y_2, in positive quadrant.)

If we are right that businessmen maximize, then right to assume only one way of producing each commodity, if no joint products, and if concerned to minimize only one factor of production. In view of assumptions, theorem does not strengthen belief in efficacy of input-output method. Each of these two assumptions less palatable than original assumption, that in maximizing state is only one activity that will be used for one commodity.

Even in long run, always more than one factor of production.

Can make slightly better case for justification of input-output by this substitution theorem if one lets some $y_1 \ldots y_n$ be inputs. Assume that specified input of specific factor has to be used. Can't put all necessary restrictions on factors in y_{n+1}, but can by putting inputs in $y_1 \ldots y_n$.

This model not per se concerned with n final products and one scarce factor. Concerned with n equalities, and one to be maximized?

In activity analysis have n inequalities, not n equalities. In input-output model of this type you are. How serious depends on the specific practical problem. If not all these factors should be scarce for some bills of goods you seek, a serious matter.

Can ask: (1) Will be scarce for all possible bills of goods? Answer likely "No." But if ask (2) Will be scarce for all bills of goods we're likely to be concerned with? May have "Yes" answer.

Another difficulty: technological change.

Not sure to what extent malign best practice by saying that deals with change by saying a becomes a'_{ij} as result of technological knowledge or extrapolation of some trend.

However may be, see what happens to substitution theorem when have technological change.

Suppose new method of producing y_2. It may lead also to new method of producing y_1, as in illustration. [Diagram with four quadrants, with y_2 on vertical axis and y_1 on horizontal axis. Two lines from first to third quadrant, through second quadrant. Left, flatter, is original pair of activities.]

All technologies may be affected as result of change in any one industry. Very hard to take account of this.

Input-output in dynamic use:

May seem would be fruitful in same way as for activity analysis. Can allow for dynamic problems of expansion associated with certain time-table of producing a bill of goods. Ability to work in a dynamic sense, taking into account inputs of capital equipment, would seem a major contribution.

But lot of difficulty. Fails to deal adequately, a/c fixed capital coefficients, and a/c difficulty in getting time lags right. Very hard to determine time might be required in future to build aircraft plant, if one is in a hurry.

Usually want more elaborate tool, that let's choose how fast want to expand capital equipment. In cases where input-output might be of real help, usually really want to make choices and take account of what is possible, not simply of what has been done. Activity analysis case is what really concerns me.

March 7: Less important matters.
(1) Mrs. Land, of London School, has paper on application of Koopman's transportation problems to coal industry:
Series of perfect inelastic supplies of coal at pitheads:
Series of perfect inelastic demands at coking furnaces.

x_{ij} takes coal from ith colliery to jth coking furnace
c_{ij} cost associated

Matrix of activity levels. Choose so all non-negative in given row add to predetermined row sum, and given sum equals predetermined column sum, with [minimization].

Here no attention paid to empty trains. That railway business.

Modify [for] bottlenecks.

Can add differing qualities of coal.

Could introduce possibility of increasing output at certain collieries at certain costs. But Coal Board influenced by other sorts of considerations – political, etc.

Empty train problem being omitted, the likelihood of uncovering non-optimal shipping practice relatively small. Mrs. Land estimates about 3%. Significant, but not spectacular.

(2) My own gimmick, with not much linear programming in it.

Listening to Colin Clark on International Comparisons of Productivity in Agriculture (output per man). Several discussants said should maximize output per <u>acre</u>, not per <u>man</u>. Example of gulf between theory and practicing statisticians.

Beginning theory says need consider <u>all</u> inputs, but little taken to heart. Probably some countries have pursued output/man to point that are over-capitalized. At least is possible. Aggregate output reduced a/c too many resources devoted to indirect production. Output/man may be higher, but output of consumer goods per man lower.

I tried to relate aggregate output to labor <u>and</u> land inputs in several countries.

Calculate ratio of output to land as well as output/labor. Plot points indicating land and labor required for one unit of output. [Diagram with land on vertical axis and labor on horizontal axis, with negatively inclined curve.]

Get rather ugly scatter. Isoquant fitted on assumption that there is a real curve, with observations deviating a/c random error. [Same diagram with scattering of x's, with curve drawn through x's along western and southern edges.]

Alternative hypothesis: Such curve exists in a perfectly efficient community. Any actual point deviates a/c less than perfectly efficient. In this case all points lie northeast of or on the curve. This estimate rapidly made.

In several dimensions also easy, introducing elementary programming.

Assumption that inefficiency sole cause of deviation. But not too realistic, a/c variation of climatic conditions. Still, can interpret isoquant as applying to most favorable climatic conditions. If not, isoquant should have some points to southwest.

Answer by generalizing model to more inputs: capital, climate, soil. But hard to measure latter. Might exclude areas of desert, mountain, etc.

Could adjust by isolating observations influenced by such factors in an extreme way.

At present computation just under way.

Obtain measure of efficiency: [Diagram with land on vertical axis and labor on horizontal axis. Line from origin through point A on isoquant to point B northeast of point A.] Ratio of actual distance to point from origin to distance to isoquant indicates efficiency. 0A/0B index of efficiency.

So far, U.K. about 67% efficient and U.S. about 55%. Rather dampens ardor of those who say British should follow U.S. But suppose difference laid to climate or fertility? Might well argue that a/c U.K. does more market gardening and dairy farming. A trial analysis on American states gave Delaware as maximum efficiency. Apparently location a crucial factor. Same result among regions of U.K. (London area leads).

Location needs to be [given] consideration as input. Need some measure of distance from large centers of population. Thinking of measure where each large city contributes amount proportional to population and distance of farm from it.

But does, in theory, give a meaningful measure of efficiency of any piece of economy.

What happens when amalgamate regions? New point is center of gravity of old. As aggregate, point moves northeast. Efficiency of point for aggregate less than that of its most efficient parts. The more you disaggregate the closer you come to the theoretical 100% efficient line, the better your estimate.

In practice, will also be random errors, which aggregation tends to cancel. If not cancelled, the disaggregated data will give curve too far to southwest. If seek most favorable conditions of climate and fertility, can argue for great disaggregation, but would you want to carry it as far as the individual farm?

Richard Goodwin, Mathematical Economics, April 28 and May 5, 1955

[Selected statements.]

1. Of the super-position theorem the notes read,

"A direct consequence of linearity."

2. Of oscillation multiplier, the notes read,

"where you have both the multiplier and accelerator working. A great generalization of the multiplier, affecting both size of income and lead or lag.

Put no lag in the system in the beginning. Everything happens immediately. Yet a lag (of autonomous expenditure on income) occurs in the system."

3. "By superposition theorem can take them one at a time. Will deal with any exogenous force whatsoever.

In most applications are interested in what system will be like if there are disturbances. Nearly everything in real world is disturbed.

As practical matter, want to know how world reacts to disturbances. Equilibrium values not important."

4. "Acceleration principle says that investment leads income by 1/4 cycle . . . . The man who got the significance of this was J. M. Clark. He said we have the explanation of the cycle. . . . In the middle of the boom the decay sets in, as investment falls off.

Hansen agreed with Clark, and told Frisch about it (at University of Minnesota). Discussion in Journal of Political Economy (1931) still good. Frisch said wasn't so. Frisch was right, as a first approximation. But lacked concept of multiplier. . . . Frisch went on to a side issue. Can <u>only</u> observe gross investment. . . . [Mathematics of replacement investment.]

<u>Essentially</u> Clark was right – had the intuition, but Frisch had the logic. Frisch had no economic intuition, and still doesn't. Frisch won . . . . Hence Clark was right if had known how to argue with Frisch."

5. Regarding the Nyquist stability criterion, "Regeneration Theory," <u>Bell System Technical Journal</u>, 1932, the notes read:

Has in part superseded the standard Routh-Hurwicz criterion. Can apply those conditions if have system and know . . . Nyquist criterion enables to see how stable the system is (as stable as it is remote from +1 point) and how to alter it to make it stable or unstable. For engineers this is very important, for must design. For economists, we want to know how system behaves, but may never know what the system really is. So design systems that behave like the real system. Important to be able to adjust systems being designed so will behave as wish.

Designing such system as want is known as Synthesis. A new field. Classical approach was Analysis.

NOTES

1. Hicks is clearly dealing with a set of tools, not a definition of reality.

2. Here the emphasis is on defining reality, which suggests that rationality and rational expectations are either tools or hypotheses.

3. Two matters are involved: (1) the empirical correctness of Malthus's analysis; and (2) the inclusion/exclusion of population from within economic models.

4. The status of the preceding two statements turns on the analyst's adherence to the neoclassical research protocol calling for unique determinate optimal equilibrium solutions, the emphasis here being on uniqueness and determinacy.

5. This and the following paragraph affirm econometrics (mathematical formalism coupled with statistics) but seemingly without the degrees of refinement and emphasis which later came into vogue – though it is hard to see how that could not have developed.

6. The emphasis on structure, as well as on taste, is unconventional but nonetheless important.

7. The use of "nihilistic" is suggestive of the strength of the hold of received theory and of apparent sensitivity to efforts to revise theory.

8. Tractability is perhaps a synonym for the desire to adhere to the neoclassical research protocol.

DOCUMENTS ON INSTITUTIONAL ECONOMICS

THE 1974 EDITOR'S REPORT OF THE JOURNAL OF ECONOMIC ISSUES

The genesis of the report published below is explained, in part, in its opening paragraph. The context in which the originating motion was passed is also relevant. The meeting in question was called to air and hopefully resolve controversy centering on the editorial policies of the journal. Juxtaposed to a policy of openness to varieties of institutionalism and of heterodox economics was a feeling by several important members of AFEE that the journal should reflect *their* conceptions of what institutionalism was all about. Accordingly, the report is, in part, a lecture, a sermon, an interpretation, as well as a record. I was very pleased that the AFEE Board of Directors supported my position, unanimously as I recall.

The document was never published in JEI while I was editor. Whether it should have been is now a moot question. After deciding in 1999 that I might publish it in an archival supplement (much material in these volumes relates to the history of institutional economics; and I still get requests for copies), I offered Anne Mayhew the opportunity to publish it in JEI. She was willing to publish an edited version but we agreed that the document should be treated as an archival matter – JEI is not an archival publication – and, moreover, the edited report is neither what I had originally written nor necessarily what I would write now.

The document is published here in its original state, with a few typographical corrections (though not the changes of "which" to "that" recommended by my spell checker). The exception is that whereas I usually reproduce the underlining in original documents, here, since the document was written by me, I have changed underlining to italics. The reader will also notice that citations in the report, which was a report on JEI, are given, in the notes, in abbreviated form.

Warren J. Samuels

THE *JOURNAL OF ECONOMIC ISSUES* AND THE PRESENT STATE OF HETERODOX ECONOMICS

Warren J. Samuels

This report has been prepared in compliance with a motion passed by the Board of Directors, Association for Evolutionary Economics, at its meeting of November 9, 1974 in New York City. The author was instructed to "write a paper that surveys what has emerged in the publication of the *Journal of Economic Issues* since its inception. That is to say, the paper is to portray the evolution of heterodox economics as reflected in the publication of the *Journal of Economic Issues*, with the paper to be submitted as a report to the Association's Board of Directors at the earliest practicable date. The minutes record as a statement of intent the view "that it would be appropriate to synthesize and assess the accomplishments of AFEE and the JEI by means of a systematic review of the JEI's contents since its inception.[1] The purpose of this report, then, is descriptive, interpretive, and, both by design and inevitably: prescriptive – as, indeed, is heterodox economics itself.[2] The aim is to be deliberately, but not excessively critical, to raise problems and issues as the basis for a research agenda rather than to affirm some version of the past. The purpose is constructive critique that is, the articulation of nature, strength, and limits.[3]

Section I will present some empirical data concerning the content of the journal through the March 1976 issue. Section II discusses and presents some conclusions concerning evaluative criteria relevant to assessing the journal's past. Section III discusses several areas of importance in interpreting the contributions of the journal. The Appendix presents an outline statement of

Documents on Modern History of Economic Thought, Volume 21-C, pages 159–189.
© 2003 Published by Elsevier Science Ltd.
ISBN: 0-7623-0998-9

the scope and central problem of institutional economics and the picture of the economy which emerges therefrom.

I. EMPIRICAL CONTENT ANALYSIS

Nineteen issues of the journal were published under previous editors and fifteen under the present editor, the latter beginning with the issue of September 1972.[4] These include those comprised both of articles accepted through the review process and proceedings and symposia. Table 1 summarizes the areas of work published in the journal by class of item and by area. Table 2 tabulates the distribution of pages by class of item. Both tables also classify by period. Tables 3 and 4 list the areas of multiple review and proceedings and symposia issues, respectively.

Several points may be noted. First, the choice of the AFEE president is instrumental regarding the topic of the annual meeting and thereby of the proceedings issue. The topic has tended to reflect the president's primary specialized field or interest. Second, articles in the issues other than proceedings and symposia reflect the flow of submitted manuscripts and the editorial selection process. Several areas of policy are relatively absent largely because of a paucity of submitted articles of quality (both quantity and quality factors operate) for example, urban economics, human resources, housing, black economic development, crime, and women. Third, published articles from the review process represent the tip of the submissions iceberg. Submissions are of highly uneven quality. Most are rejected due to their lack of a substantive contribution or inadequacy with regard to their intended purpose or field. Submissions have been somewhat unstable between 1971 and 1976, ranging between 160 and 200 manuscripts a year. Approximately 5 to 7% have been accepted.

The distribution given in Table 1 should not surprise anyone familiar with the journal, especially the leading dozen or so areas. There are differences between the two periods which will interest alarm, and/or please different readers. It should come as no surprise that the sum of the following areas represents 15% of the total publishing activity of the journal: industrial organization (which includes antitrust), economic role of government public utility regulation, capitalism, the corporate state/system, and social control of business. The proportion of proceedings and symposia issues devoted to these subjects is substantially greater. Tables 1, 3, and 4 also reveal, not unexpectedly, considerable publishing activity during both periods in such areas as the economy as a system of power, the institutional foundations of the economic system, the design and performance of institutions, and the economic role of government. Finally it may be noted from Table 2 that the average per-issue page size has doubled between the two periods.

II. EVALUATION: SOME GENERAL CONCLUSIONS

The judgment of success or failure in part depends upon the definition of identity and/or criterion(ia) of purpose employed. There is a tension between, inter alia, making impossible demands, or creating or pursuing unrealistic expectations, and relying upon comfortable diffuseness. There is also the question as to how much (by whatever criterion) one journal can accomplish in nine years vis-à-vis both the profession (if not also society) as a whole and the number of active scholars in its field of inquiry. This, too, depends upon expectations and criteria of evaluation. Above all are conflicts as to identity and purpose.

This section presents the author's evaluation with regard to, first, past statements of purpose (including his own) and, second, other relevant criteria. Whatever are the merits of the former, they do constitute one not unreasonable and irrelevant basis for judgment.

Past Statements of Purpose

The Constitution of the Association for Evolutionary Economics, as amended in 1970, states in Article II that the purpose and objectives of the Association shall be to foster, in the broadest manner, the development of economic study and of economics as a social science based on the complex interrelationships of man and society and in a manner such that will acknowledge the need to join questions of economic theory to questions of economic policy. In the first issue of the journal, Forest G. Hill, Editor, wrote:

> The title of this *Journal* may suggest its main emphasis. It will be broadly concerned with major issues of public policy, of economic methodology, and of understanding the processes and problems of economic change. This general scope includes issues which have been made prominent by institutional economics and social economics as well as by current economic changes. These issues are typically interdisciplinary in character; they pose problems of the relation between economic theory and economic policy; and they often bear on the changing role of government in the economy.[5]

Speaking of the articles in that issue, Hill wrote that "they illustrate fairly well what we have in mind for the scope and emphasis of the *Journal*. Without doubt, most of these articles raise policy issues and methodological questions. Most of them also deal with interdisciplinary aspects and with values or goals. Interestingly too, most of them pay some attention to non-economic factors in economic change and to the development of economic thought."[6] In the

Table 1. Distribution of Articles, Notes and Communications, and Book Reviews.

	Articles, Notes and Communications				Book Reviews		Totals			
	Period 1		Period 2		Period 1	Period 2				
	Regular Issues	Prcdgs & Symposia	Regular Issues	Prcdgs & Symposia			Regular Issues	Prcdgs & Symposia	Book Reviews	Total
Growth and development	29	2	3	6	32	19	32	8	51	91
Analysis of institutions	32	10	5	5		5	37	15	5	57
History of economic thought	19	2	15	5	8	7	34	7	15	56
Economic role of government	10	4	3	20	6	6	13	24	12	49
Institutional theory	10	8	1	12	4	9	11	20	13	44
International trade and finance	13	4		7	9	10	13	11	19	43
Comparative economic systems	7	3	5	2	11	12	12	5	23	40
Macro-monetary	7	3	6	12	4	5	13	15	9	37
Industrial organization,	3		3	22	3	5	6	22	8	36
Labor economics	10	4	2	6	11	1	12	10	12	34
Economic methodology	9	1	8	6	3	2	17	7	5	29
Institutional economics	5	2		13		3	5	15	3	23
Critique of orthodoxy	8	2	1	8		1	9	10	1	20
Income distribution-poverty	6	1	1	2	8	2	7	3	10	20
Marxian and radical economics	3	1	5	1	1	9	8	2	10	20
Environmental economics		2	4	6		7	4	8	7	19
General economics (misc.)	3	3	3		4	6	6	3	10	19
Public finance	2		5	1	6	4	7	1	10	18
Economic history	2		1	4	4	5	3	4	9	16
Welfare economics	4		5	4	4	3	9	4	3	16
Corporate state/system		5	2	8		1	2	13	1	16
Public utility regulation	6	1		4	1	3	6	5	4	15
Social security and welfare	4				7	3	4		10	14

Table 1. Continued.

	Articles, Notes and Communications				Book Reviews		Totals			
	Period 1		Period 2		Period 1	Period 2				
	Regular Issues	Prcdgs & Symposia	Regular Issues	Prcdgs & Symposia			Regular Issues	Prcdgs & Symposia	Book Reviews	Total
Economic philosophy	2	1	2	3	5	3	4	4	6	14
Technology	3	1	3	2	1	4	6	3	5	14
Consumption theory			1	13			1	13		14
Economics of education	5	1	1	1	3	2	6	2	5	13
Economics of medical care	4		1	2	3	2	5	2	5	12
Economic planning	1	1		3	2	1	1	4	3	8
Economic anthropology	5				2		5		2	7
Agricultural economics		1		1	1	4		2	5	7
Capitalism		6				1		6	1	7
Black economic development	1			1		3	1	1	3	5
Social control of business				5				5	5	5
Human resources		2				1	1	2	2	4
Economics of housing						4			4	4
Economics of women	1	1					2			2
Natural resources					1		2			2
Population		1						1		1
Economics of crime	1						1			1
Urban economics						1			1	1

Note: Period 1 through March 1972. Period 2, September 1972 through March 1976. Some items represented in more than one category.

Table 2. Distribution of Pages.

	March 1967 – March 1972 (19 issues)		September 1972 – March 1976 (15 issues)	
	Pages	Percentage	Pages	Percentage
Articles	1785	85	2441	80
Notes and Comments	118	6	94	3
Multiple Reviews			208	7
Single Reviews	199	9	300	10
Total	2102	100	3043	100

September 1969 issue, the new editor, Harvey H. Segal, wrote that the journal "will continue to concentrate upon the broad issues and problems of economic policy. It will continue to open its pages to ideas and styles of analysis that seldom receive congenial receptions in other journals."[7] Finally as part of his first Editor's report, to the December 1971 AFEE membership meeting, the present editor outlined the following as editorial policy:

(a) a journal open to all shades of opinion, a scholarly journal, aiming at quality substantive contributions to fields appropriate to the journal.
(b) a wide range of articles, chosen from submissions.
(c) promote development of corpus of institutional analysis as body of knowledge.
    (i) anticipate a healthy tension between this and other goals.

Table 3. Distribution of Multiple Reviews.

1.	Analysis of institutions	11
2.	Economic philosophy	8
3.	Comparative economic systems	4
4.	Institutional economics	3
5.	Economic growth	2
6.	Labor	2
7.	Methodology	2
8.	Psychology	2
9.	Radical economics	2
10.	Income distribution	1
11.	Industrial organization	1
12.	Law and economics	1
13.	Welfare policy	1

Note: Some items represented in more than one category.

Table 4. Distribution of Theme Proceedings and Symposium Issues.

	Theme Proceedings	Symposia Issues
Capitalism	1	
Chicago school		2
Consumption theory	1	
Corporate state	1	
Law and economics		1
Macroeconomic institutional innovation		1
Market, institutions and technology		1
Social control of business	2	

These statements, at least one of which represents the result of compromise among individuals with different conceptions and purposes, are diffuse and, taken singly or together, not without internal conflict and ambiguities. Such is inevitable. Nonetheless, they do convey a general focus for the journal. It is this focus, only imperfectly articulated in the various themes encapsulated in the foregoing statements, which the journal, I think, has, from its very beginning provided. The JEI has fulfilled its broad mandates to be eclectic, holistic, evolutionary, policy oriented, and institutionalist-heterodox, while remaining open, say, in article authorship, books reviewed, and book reviewers, as well as membership on its editorial board, to those closer to being followers of neoclassicism and Marxism, yet sympathetic to the interests of the journal. There is considerable diversity of approach, interest, and position among the contributors to the first nine-plus volumes, a diversity which may be disturbing to some inasmuch as it betokens a lack of overriding central aim; but overall, and certainly in comparison with the more mainstream economics journals (which for the most part are not by any means closed to heterodox thought), the JEI has been open to and thereby promoted eclectic holistic, evolutionary, policy oriented, and institutionalist-heterodox thought. In this it has been satisfying the goals of those who have avidly supported it and rendering an important service to the profession.

No less important, however, is the question of the quality and/or magnitude of the fulfillment of the general goals. A number of issues may be considered in connection with specific evaluative criteria.

Some Specific Evaluative Criteria

Institutional economics, if not heterodox economics more broadly contemplated, has been multifaceted. It has comprised: (1) a protest movement against capitalism and/or market economics, or their excesses; (2) a problem solving approach: and (3) an attempt to develop a body of knowledge. To some extent these directions have been reinforcing, but they also have conflicted, as different individuals have emphasized or been preoccupied with different negative and/or affirmative beliefs. It is almost certain that the journal could not have satisfied completely the desires of individuals with these diverse goals. Certain aspects of institutionalist preachment and practice will be considered in this section in regard to evaluative criteria; others, dealing with more substantive questions, will be examined in the following section.

(1) *Critique of Existing Market Economy and/or Market Economics*

The journal certainly has provided an outlet for considered critiques of both types. The critiques have been diverse, reflecting the heterogeneous character of contemporary economic heterodoxy as well as of both the market economy and market economics. Some writers have rejected the market system per se. Others criticize only the existing market system; these, and others, desire an improved and effective market system. There has been great variety among the critiques and among the preferred alternatives. Similarly with the critique of market economics. Some have rejected neoclassical economics. Others reject only its excesses and urge attention be given to its limits; these, and others are willing, even avidly, to use it where applicable to their research interests. With a few exceptions, and for better or worse, the JEI has lacked the bitter, acrimonious, and uncompromising rejection of orthodoxy; it often has demonstrated eclecticism, openness and holism. Heterodoxy includes both the alienated and unalienated. The general tone of the journal, however, has been constructive and not hostile, even when the criticism has run very deep and/or been severely delivered.

(2) *A Problem Solving Approach*

Many institutional-heterodox economists have been problem or policy oriented in both their interests and activities: many others have not been as active in the worldly arena, sharing interests and orientations but preferring the relative quiet of the contemplative life. To both groups, perhaps in unequal degree in practice, however, institutional economics has represented a multidisciplinary approach to problem solving, most notably of questions of public policy and reform. This approach has comprised a search, wherever knowledge and insight might be found for means to chosen ends and as such has represented a blending of normativism with regard to ends and positivism with regard to means (a condition characteristic of many orthodox economists, such as the Chicago School). This approach often has been characterized as either instrumentalism or pragmatism, following such writers as Lester Frank Ward, Adolph Lowe, John Dewey, and Clarence Ayres.[8]

Central to this effort is informed design and performance evaluation and prediction of institutions. Articles and other materials published in the JEI have made contributions along these lines although no distinctive style or strategy has been developed in technical detail. Encouragement of such work remains an important, however difficult, task for the future.

A problem solving approach, even if primarily of a general interdisciplinary orientation, is valuable for purposes other than problem solving. It is or can be knowledge producing. It can open up the end-determination or valuational process. It can harbor reformism in a constructive manner. Finally, and no insignificant matter in my view, it can project the economy itself as a process of solving a succession of problems.[9] There are limits, however; for example: The question is, rather, whether we gain useful insights or mislead ourselves and others when we use the tools of our trade as citizens participating in the political and governmental processes through which legal rules are developed and changed.[10] However well intentioned and careful the search for problem solutions, there is no guarantee of performance success, not to say adoption. Indeed, the complex set of factors which produces performance results, including those which govern political adoption of institutional innovation, should be among the self-conscious concerns of the type of institutional economics outlined below – including the pluralistic nature of the total system,

(3) *Advocacy and/or Realization of a Specific Policy Orientation and/or Program*

Many institutional-heterodox economists desire more than critique and a problem solving approach. They prefer advocacy of particular policy orientation and/or a specific policy program. Although the AFEE Constitution is considerably more restrained,[11] many desire to promote purposeful social change in particular directions, urging a program of political activism, with AFEE as such remaining a learned society. Many others prefer not to prescribe policy goals, although they may have their own distinctive orientation and preferences. Nevertheless, there is some belief that the institutionalist tradition is unduly narrowed by concentration upon descriptive work.

There are three points to be made. First, institutional economists generally have been sympathetic to twentieth century liberal reformism in favor of the social control of business, worker interests in a pluralistic capitalism, and, in their own conceptions, a viable economic system. Yet, second, there has been and is great diversity within institutionalism, as within economic heterodoxy generally. There are substantial differences if not conflicts over the market, planning, and, inter alia, the desired means to the aforementioned reformist ends. Third, the pages of this journal have not been the vehicle of any policy program or specific orientation. While generally resonant with liberal reformism (but including both those who want to constrain and to develop the market, for example), its pages have been open to all comers, subject of course, to the editorial requirement that each manuscript rake a substantive contribution.

(4) A Body of Knowledge

The criterion most unequivocally appropriate to a scholarly journal is that it publish contributions to and thereby promote the development of a corpus of knowledge relevant to its field of interest. The main role of the JEI has been to provide an opportunity for publishing substantive contributions to institutional-heterodox economics, material which is critical and/or constructive. The JEI has published contributions to the development of economic knowledge, particularly the creation and expansion of a body of institutional analysis. These contributions have been to both paradigm development and substantive analysis. Within its pages are empirical and theoretical contributions which may be seen as either alternative or supplementary to the corpus of mainstream economics.

The question remains as to how viable as research fields are the basic problem areas and formulations undertaken and advanced by institutional economics, say, in the pages of the JEI. In my view, the JEI demonstrates the meaningfulness of and great opportunity for research in holistic and evolutionary economics. That such meaningfulness and opportunity exists within the pages of the JEI however much it awaits systematic integration, is the major conclusion which I offer in this report. Devotees should be less concerned with strategic or evaluative issues and more productive of research and publications.

Apropos the question of viability, it should be clear that institutionalists, that is, those who seek to make contributions in the spirit of the focus of this journal, have taken on difficult if not intractable and relatively unsolvable problems, but problems no less important for being so.[12] Neoclassical economics has the analytic (and to some extent normative) advantage of accepting, working within, and promoting the market and institutions enabling the market. Marxism has an advantage partially derived from its philosophy of history. Institutional economics, particularly insofar as it is an attempt at descriptive, explanatory, and interpretive knowledge, is much more ambitious. Often, in part to avoid merely affirming the optimality of the status quo, institutionalists have to work at a highly abstract level not readily yielding operationally researchable hypotheses (especially in quantitative terms), lacking a so-called hard cutting edge, and not directly productive of policy inferences, nonetheless producing considerable historical *and* analytical insight.[13] Institutional economics has tended to specialize in important problem areas not or not yet, if ever, easily or meaningfully treated operationally. That fact is both its burden and its badge of distinction. Whatever else can be said of the work of the institutional economists, the problems with which they typically choose to work are neither trivial nor formulated in a trivial manner,

Neither this journal nor institutional-heterodox economics as a whole (with the exception of Marxism) has achieved an integrated, coherent corpus of analysis. There are concepts, lines of reasoning: problem foci, and accumulated substantive material, but as critics both friendly and unfriendly have pointed out over the years, there is no body of integrated theory such as characterizes neoclassical and Marxian economics and which so readily reduces to textbook treatment and interstitial research by graduate students. Although the absence of such a corpus is to some extent a function of its holistic and eclectic orientation, it is an important shortcoming but not an unequivocal one. It does have to be recognized, however, that institutionalism may have no elaborate corpus of analysis, rather only a way of investigation with some important guides to inquiry. I think that institutional economics does involve more than this but am prepared to recognize that substantively it may be closer to an investigatory mode than a corpus of knowledge comparable to other schools of thought.

The creation of such an integrated, coherent corpus of analysis requires, in part, a clearly outlined orientation, a paradigm and/or a subject matter defined by a meaningful central problem(s). A great deal of paradigm-level work has been done, including material published in this journal, but, alas both institutional-heterodox economics and this journal lack an explicit and widely shared and developed paradigm or cognitive system except in the most general terms for example holistic systemic, evolutionary. The dismay indicated in the immediately preceding sentence would not be shared by all, however. To many, the attraction of institutional economics, and of the JEI, is its eclecticism, which might be impaired by the sharing of an explicit paradigm, not to say a coherent corpus of knowledge. Indeed, the absence of a strong paradigm may constitute the strength of heterodoxy. Yet, such a view has its limits.

The JEI has published not only substantive material but also discussions which comprise the germs of one or, indeed, more such paradigmatic formulations, perhaps even more than mere germs or hints But if both the formulation of a corpus of institutional analysis and the articulation of an acceptable and meaningful paradigm are to occur, much clarification and integration remain to be done. In the Appendix, I present one attempt to formulate the scope and central problem, that is, the paradigm, characterizing the field of institutional economics and the picture of the economy which emerges from it.

It should be said, too, that the JEI has received and published little creative work directly expanding upon the historic contributions of Thorstein Veblen, John R. Commons, and Clarence Ayres, among others, and upon the contemporary contributions of John Kenneth Galbraith, Gunnar Myrdal, K. W. Kapp, and Kenneth Boulding, among others. There has been some explication and exegesis, but there have been submitted too many attempts at retrospection and too few at analytical development. The need is to use and transcend and not merely to rehearse the old.[14]

There are, of course, limits to the degree of integration which a journal as such can achieve in its field of interest and inquiry. Both institutional-heterodox economics and the JEI can be a focus of meaningful economic thinking (positive and normative) *without* either a corpus of theory or a shared paradigm. Yet, it would be quite useful to have them.

It is obvious that neither institutional-heterodox economics nor this journal has produced a major shift in economics. Neither have brought about significant changes in the way people think about the economy, although (for example) the work done by Galbraith and others has increased sensitivity to the phenomena of the corporate state-system, the result abetted by the pressure of reality itself. If the criterion of success be the production of a major shift in economics, then the journal has failed. Such a criterion, however, may be rather unrealistic. At the other extreme, however, the journal should not be merely the haven for the private prejudices of a few malcontents talking only to themselves.

The journal has contributed, I think, to the respectability of institutional-heterodox economics beyond Marxism, and this is no mean feat. Then, too, there is no minor amount of sympathy among more nominally mainstream economists with the interests of and issues raised by the journal and its contents. The dominant systems of incentive and reward and of professional training operate to inhibit major shifts within the profession. But clearly there is a place for institutional-heterodox economics and this journal. Were that only translated into increased subscriptions!

Within heterodoxy, of course, Marxian economics is a powerful rival, as it were, but with much supplementary material. In addition, orthodox economists have worked in and contributed to the fields of interest to this journal, about which more in Section III.

One can say, too, that the journal has enabled the survival of institutional economics among a self-conscious group with a sense of identity, however amorphous. This is a main function of a journal such as the JEI, and for the most part it seems to have rendered such a contribution.

Survival is important; as Aristotle argued, it may be the key test. Less dramatically, the JEI has served as an outlet for heterodoxy, including institutionalism of the traditional variety(ies). This, too, is a main function and contribution. In so doing, among other things the journal has encouraged and abetted promising young scholars active in the journal's field of interest and inquiry and interested in its approach. The JEI makes a contribution by being a heterodox Journal *of quality.*

The JEI is not an end in itself, of course: yet it does play critical roles. It is one voice of heterodoxy in Western economics. Indeed, if heterodox economics were to come to dominate the profession, we would require the critiquing of heterodoxy itself and of course internal critiquing already exists,[15] heterodox economists being no less critical among themselves than economics generally.

III. SOME FURTHER CONSIDERATION

This section explores further aspects of JEI evaluation, interpretation and prescription. It first relates the critique of orthodoxy to the limits of heterodoxy itself. Second, it raises the problem of positive vs. normative institutionalism (and heterodoxy in general) in the light of the heterogeneity of heterodoxy and the foregoing limits. Third, it considers the problem of methodological (epistemological) differences between neoclassical and institutional economics. Finally, it examines the relations of institutional to neoclassical and Marxian economics.

On the Critique of Orthodoxy

In my view, the highest level of meaning in economic thought is on the level of the matrix formed by alternative theories and schools and their mutual critiques.[16] In this context, and treating the matter very limitedly and generally, the role of heterodoxy to orthodoxy is determined in part because the latter *is* orthodoxy and in part because of orthodoxy's frequent close connections with the status quo in a rationalizing role. In both respects, heterodoxy exists in relation to and in a sense dependent upon orthodoxy's tendency toward artificial or contrived absolutism and closure, for example, in elevating orthodoxy's ostensible determinate solutions as pre-eminent to the actual radical indeterminacy of reality. Heterodoxy's critique is thus a function of the general and specific strengths and limits of orthodoxy per se. Moreover, insofar as heterodoxy is heterogeneous, its critique will be likewise and insofar as orthodoxy is heterogeneous, heterodoxy's critique will be further heterogeneous.

Apropos the critiquing *role*, I would say first that there are both vulgar and non-vulgar formulations of orthodoxy and of heterodox critique, Second, the critiquing function does not have to be conducted in a shrill and alienated manner. It can be a valuable intellectual exercise and disciplinary contribution both as critique and as the basis of alternative conceptions of economic reality.

The JE1 reflects and/or has made contributions to the foregoing. The substance of the critique of neoclassical orthodoxy found in the journal has been of two general types: one, pointing out facets of orthodoxy which contrast with its idealized self-image the other, critiquing from an extra-paradigmatic orientation the content and limits of orthodox thought. In both respects, the critique has been heterogeneous, for the aforementioned reasons. It will be useful to summarize the main thrusts of the two types.

With regard to the infirmities of orthodoxy's idealized self-image JEI writers, first, have emphasized the relativist character of the development of mainstream economic thought Its evolution has been a product of ideology and strategic consideration, and the boundaries of neoclassicism have been a matter of taste. Second, attention has been called to the ambiguity and heterogeneity of neo-classicism. It is at once a theory of the market, choice, resource allocation, and methodological individualism. Moreover, there are enormous gaps between the pretensions of mainstream theory and our understanding of the economy. Third, neoclassicism has been a kaleidoscopic mixture of explanation and rationalization of the market system: It has been a blend of positivism and normativism. Its ostensible positive analysis often if not always contains deep and selective normative choices and biases for example, selectivity as to which (whose) preferences are to count, and it is quite thoroughly imbued with values. It has practiced presumptive optimality reasoning, thereby giving (often inadvertently) effect to (varying) implicit antecedent normative premises and elastic technical assumptions. It has served, moreover, a high priest role, shielding the system from criticism and obscuring the scientific inadequacies and limits of neo-classical analysis in order to preserve its identity and ideological role. It is as much, if not more, a system of belief as a body of scientific tools and theorems. In all this, heterodox writings in the JEI (and elsewhere) have performed a demythicizing role. JEI authors also have pointed out the irony of limited neoclassical techniques, models, and theorems juxtaposed to the frequent and profound recognition by neoclassicists of broader and deeper considerations governing the organization and operation of the market. (The irony also applies to institutional analysis itself; see below.)

Heterodox-institutionalist critique of the substance of mainstream economics has been directed, first at the narrowness of neoclassical analysis: Its range of variables and length of causal chains have been unduly truncated. However defined, its central problem has taken a narrow view of the economy and thereby dealt only with a slice of real-world radical indeterminacy. It has tended to adopt a posture of systemic inevitability rather than fecundity. It has operated within its own paradigm and posed and answered questions only on its own terms such that its conclusions are tautological with its systemic and other assumptions. Its focus is upon partial equilibrium models and perhaps thereby, has a tendency toward single-factor explanations in a general interdependent world. Its methodological individualism is severely limited by real-world methodological collectivist processes and, indeed by methodological collectivist elements obscurely but importantly encapsulated (often in truncated form) within its analysis. It tends to neglect or narrowly define the organization and control fundamentals of the economic system; the market is conceptualized as a producer of order rather than also as a facet of a larger formulation of order and as a product thereof as well. Its methodological individualism is limited and selective as well as generally independent of important social forces, processes, and problems. Its excessive and/or naive rationalism-hedonism is often extended into a vision of the economy and society as a whole. As policy analysis among other things, complex institutional and historical situations are often forced into the neat and logical categories of price and market theory, sometimes with serious deficiencies or with limits typically forgotten. It tends to de-emphasize obscure, and even ridicule ("that's sociology") important problems either not amenable to the tools of economic analysis or treated only incompletely thereby, more often treated presumptively through taking as given certain problem solutions and not others. Its analyses tend to be primarily short run. It artificially (and conservatively) separates production and distribution. Neoclassicism correctly portrays a general system of tendencies, yet its affinity for single-factor analysis tends to produce inferences which tend to be unidirectional when in fact, under certain conditions, they may be found reversed.

A second and closely related line of critique has pointed to neoclassicism's neglect of considerations of power, conflict, distribution social relations, non-market institutions and processes, and/or their interpretation only within the limited confines of the neoclassical paradigm and its subtle (and sometimes heavy handed) emphasis upon harmony, equilibrium, and optimality of markets. Mainstream theory has selectively treated such phenomena as freedom, power, externalities government, and coercion and has tended to neglect the non-uniqueness of Pareto optimal solutions. It has neglected the factors and forces governing the distribution of welfare, for example, those governing the simultaneous determination of output and distributional variables (including product definitions). It has generally failed to differentiate the actual structure of economic power from the general conception of the market system as alternative modes of representing economic reality, in part failing to differentiate adequately between competition and the market per se. It has similarly failed to acknowledge the profound structural changes that have taken place within the system" and perhaps *of* the system. It has failed to comprehend the price and market system as an institution or set of institutions and how such phenomena as the price level, prices and wage rates, and the supply of money are institutional variables and phenomena and how, in some ultimate sense, they are influenced or governed by the structure of (and changes in) institutions, that is, the distribution of power. It often has failed to deal with the conflicts produced by economic development and, accordingly, with the social functions of conflict.

Third, heterodox critiques have been directed at the selectivity with which neoclassicism treats the role of choice in human society. Its mechanistic and formalist techniques – its solution mystique – have introduced a determinism into economic analysis vis-à-vis the deep choices and choice processes (themselves subject to choice and thereby in competition) in society. It thus has aborted a meaningful understanding not only of the general interdependent character of the economy but also of the role of volition and the fact of powerful radical indeterminacy therein. Not only does it tend to be unilinear in its characterization of the operation of the system, but also its mechanism and selective narrowness permit if not generate the nuances and implications of laissez-faire and thereby readily absorb traditional conservative prejudices and promote and give effect to the status quo power structure (or selected facets thereof). In reality that power structure itself is open to choice and revision, and neoclassicism thus becomes a participant in the very system it purports to observe and study. Neoclassical theory inevitably involves tautologies given selective and/or ex post specification which may vary from those found in reality or which give presumptive status to selected performance results as if inevitable. Along a related line, JEI authors have pointed to neoclassicism's focus upon the market not only as being culture bound, but also as being a focus upon one facet of the economic system in a way which treats the economy as self-contained and autonomous. In reality the economy: (1) is a dynamic, both determined and determining cultural phenomenon or complex; and (2) social factors are subject to change through human action. Neoclassicism thus has tended to a reductive reification of data and processes; mechanistic determinism obscures both cultural evolution and the fundamental volitional character of man and the economy. Along a related line, neoclassicism is held to concentrate attention upon the market as the context for the development of man with economic man either an institutional assumption or a tautology given selective formulation; it fails to give systematic consideration to man as both a product and producer of social forces other than those phenomena encapsulated within the market or seen as regulated by the market. These social forces are important even if market-like mechanisms (for example, constrained maximization) are present therein. Neoclassical prediction is limited by cultural evolution, novelty and volition. In general, it is constrained by its neglect of process (for example, systemic change), its ahistoricism, its cultural specificity and its limited appreciation of the fact that economic phenomena are a function of artificial human selection through the exercise of will or volition and not the working out of natural forces, notwithstanding the important regularities which may be found and charted. Neoclassicism thus tends to neglect joint determination processes; for example, by focusing upon the conditions of optimality it tends either (if often only implicitly) to prescribe or neglect the interrelated processes of rights determination and redetermination which govern the substance of optimal solutions in the real world and in theory. Running throughout all the foregoing is the neglect of: (1) worker interests in the definition of output and of work as a meaningful activity in and of itself; and (2) the forces governing the formation of individual preferences

Finally, among other things, JEI authors have critiqued neoclassicism's treatment of the status quo: The status quo is often incompletely and/or ambiguously defined or specified. Nonetheless, neoclassical analysis and policy recommendations tend to be system and status quo specific. Thus, neoclassicism is both ambiguous in its applicability to the real world and discriminating in its reinforcement of the status quo through selective attribution of propriety and impropriety and through policy inferences whose effectuation would not be Pareto-better. Neoclassical analysis is further ambiguous and selective in its application to the changing status quo. Finally, the role of heterodoxinstitutionalist critique has been to point out the normative elements taken for granted and given effect by neoclassical analysis, for example the structures of institutions and rights.[17]

On the Limits of Heterodox-Institutional Economics

As already indicated, the foregoing critiques may be found in both vulgar and non-vulgar form. Too often a critique is formulated in a manner at least as naive as the crudest formulation or use of orthodoxy. Caricature is found on both sides. That is an important limit to heterodox critique, however much it is countered by the extravagances of neoclassical epigones and ideologues. However, the main point I wish to make is that the foregoing heterodox critiques of orthodoxy also apply, *mutatis mutandis*, to heterodoxy itself. Institutionalist and heterodox economists have no monopoly of analytical purity, completeness and normative virtue. Their work shows myopia and selectivity no less than that of neoclassicism. They have their own distinctive balance(s) of specificity and generality. Their work has limits imposed by virtue of operating within their own paradigmatic and conceptual formulations. Emphasis upon holism, evolution, process, and eclecticism does not save institutionalists from committing the same or similar questionable practices as their neoclassical counterparts, not the least being single-factor explanations. Institutional and heterodox economists have produced bodies of thought which are as relativist in their development, as ambiguous in their central meaning, and as much a blend of explanation and rationalization and of narrowness and breadth as neoclassicism. Institutional and heterodox economists have focused upon and given their own distinctive twists to a narrow set of variables. They too have treated the role of choice and volition in human society quite selectively. They too have been both ambiguous and selective in their treatment of the status quo. They too have not fully dealt with considerations of power, conflict, and so on. Institutionalist and other heterodox economists not only have performed the role of high priest but also have introduced their awn forms of closure, determinism, and presumptive propriety reasonings. And, of course, substantive institutional research and writing is subject to its own analytical and other limits. All of this is further complicated (both affirmatively and negatively) by the absence of a specific institutionalist paradigm and tight corpus of knowledge, as paradoxical as that may be.

Amplifying somewhat upon the above, first, heterodox economists tend to neglect the fact that explanation-analysis and defense of the market can be a ceremonial justification of the status quo power structure but that (in part as such) it can be an instrumental justification of a means of enhancing the life process through technological innovation and/or the diffusion of power.

Second, apropos, for example, of the ceremonial-institutional vs. instrumental-technological distinction, some fundamental institutionalist differentiations are ambiguous and subject to selective normative specification. Emphasis upon technological imperatives and the logic of industrialization, moreover, tend to absorb institutionalism itself in culture bound, single-factor, and non-volitional mode of analysis.

Third, it must be recognized that institutionalist writers are motivated by and infused with their own motivations and that heterodoxy is likely to have its own dogmas and preconceptions of the past and present.

Fourth, it is uncertain as to the degree to which the holistic conception of the economy can be: (1) specified; (2) separated from the rest of society; and (3) made manageable for analytical purposes, quite aside from its being made operational for testing purposes. There is enormous room for interpretive flexibility in the study of the large problems with which institutionalism deals: this is both a limit and an opportunity, of course. On this I would add the following. There is much partial-equilibrium (or partial-interdependent) or single-factor analysis within institutionalism which has (no less than neoclassicism) the problem of overdoing the fruits of abstraction. There is some reification of social forces and consequent ambiguous treatment of the volitional element. It is difficult to isolate the effects of certain forces, for example, power, from other influences. There are complex interpretive problems in assessing the historical role and extant normative status of institutions, and so on. In short, there are enormous gaps between the pretensions of and actual understanding provided by institutionalism.

One can say also that the identity of both institutionalism and heterodoxy in general is very much, perhaps too much, defined by the orthodoxy to which they react. They are the bull to the latter's red flag. Institutionalism and heterodoxy should come to be defined less by their critique of orthodoxy and more by their affirmative constructive work, even if only in their practitioners' own minds. Institutional economics may well be limited to a pariah role in the profession for several reasons: It is heterodoxy per se; it does not take the system in its present form or state as given but studies it as a variable; and since the dominant economics in any society reflects the status quo system and power structure (although in no close one to one relationship), institutionalist-heterodox economics is not likely to become central in any market or socialist economy,

I want next to consider two problems which run through the foregoing, namely, the heterogeneity of heterodoxy and the positive-cum-normative character of institutional economics. I will then consider the problem of methodology and finally the relation of institutional economics to neoclassical and Marxian economics within the larger discipline.

The Heterogeneity of Heterodoxy

The limits of institutionalism-heterodoxy discussed above are evident in the pages of the JE1 especially in regard to its heterogeneous composition. The fact is that heterodox economists differ among themselves on such fundamental questions as what the economy is all about, how the economy is in fact organized and controlled, and how it should be changed. Not only is institutionalism fractured by virtue of its being a reform and protest movement, a problem-solving approach, and an attempt at a body of scholarly knowledge, but also there are multiple foci of emphasis and different approaches to each one. This heterogeneity is both a limit and a saving grace. Each institutionalist must work out the heterogeneity individually, which is much less of a problem for the fledgling (and mature) neoclassicist, who has a relatively coherent inherited paradigm and corpus of knowledge within which to work. In part, this is the price of eclecticism, and in part it results from institutionalism-heterodoxy being the haven for those unhappy with a rigid intellectual framework.

Institutionalist-heterodox economists have a truly wide range of interests and competing emphases: technology, power institutions, the state of the working class, instrumentalism, pragmatism conflict and its resolution, social forces, distribution, evolution, philosophical-ethical relativism or absolutism, progress, contextualism, economic organization and control (including the differentiation between price-competitive and organized-planned sectors) deliberative social control and/or social change, institutional design and performance, socialization of the corporate system, economic planning, the economic role of government, the logic of reform and/or industrialization, humanism, and, inter alia, economic development and growth. There are alternative formulations of problems and issues, different visions of holism and evolution, different areas of specialized interests, and reformist urge without agreement as to an agenda.

The burden of this heterogeneity is that the JEI likely will not satisfy any partisan of a particular interest or orientation if it is to be open to all, upon the condition of substantive contributions. The burden of the other limits of institutionalism-heterodoxy for the journal is also obvious: Journal articles, especially more so than treatises, are particularly subject to the problems cited or alluded to above. The JE1 in fact reflects these problems all the more conspicuously because they are presented in black and white. The multiple reviews of Allan Gruchy's *Contemporary Economic Thought*[18] raised serious questions concerning the professional status and the nature and substance of institutional economics. These are not unique problems but issues which were bound to be raised anew by Gruchy's effort to interpret contemporary (neo)institutionalist thought. One of the issues confronting the *Journal of Economic Issues* is the nature and substance of institutional economics itself.

Positive and Normative Institutional Economics

The JEI reflects the heterogeneity of institutional-heterodox economics in another way: Its contents reflect varying blends of positive (meaning description and non-judgmental interpretation) and normative (meaning evaluative and prescriptive-proscriptive) work. There seems to be general agreement that values inevitably enter into positive work and that the best solution (at least as a beginning) is that of Gunnar Myrdal, namely, their specific iteration; accordingly I shall say nothing more on that.[19] More important for present purposes is whether, that issue aside, institutional economics should comprise positive or normative work, or both. There are those who feel that institutionalism should concentrate upon positive, non-normative description and interpretation of the factors and forces governing the evolution, organization, and operation of the economy and not indulge in passing judgment thereon, whatever the private prejudices of practitioners. On the other hand, there are those who feel that confining institutionalist analysis to description (even including interpretation) is too narrow a conception, that institutionalists should not abandon the field of values and evaluation to politicians, bureaucrats, economic interest groups, and other economists.

In my view, just as there should be several schools of economic thought, including institutionalism, institutionalism should be capable of including both positive and normative work. Institutionalism will best make its way in the world if it strengthens its positive contributions as distinct from its normative. But, more broadly, it is I think, one of the beliefs of institutionalists that in the real world (and institutionalism is part of the real world) there is a general interdependence of positive and normative endeavors; certainly this is central to the theory of pragmatism which many find suitable for institutionalist work. By the same token, positive institutional economics should be capable of including both liberal and conservative normative analysis of the development, role, and operation of institutions, and so on. So, too, should there be developed skills and knowledge relating to the design, performance prediction and analysis of institutional innovations, say, with any evaluative criteria clearly specified.

Regardless of my own views, however, there is considerable manifest and latent conflict within the pages of the JEI and the larger movements of thought which it somewhat records. There is tension between normative and positive and between various specifications of each. There is a great deal of so-called analysis which is but a euphemism for the writer's preferences. There is both criticism of others' myths and the promulgation of one's own. In all this, the proposition that evaluation is a function of perspective, which one might think would be a notable point in institutionalist analysis, is often neglected by institutional economists operating within their own perspective.

In my view, it is both possible and desirable for institutional economists carefully to develop both positive and normative bodies of analysis. This has been and should be a basic objective of the JEI.

The foregoing does not intend to de-emphasize the considerable heterogeneity which marks the institutionalists' positive analysis. Nor does it attempt to obscure the even more marked differences among institutionalists on normative issues: they are both for and against the market, particular changes, regulation or deregulation, socialism, planning, reform of the market and/or government, and whether more is better. Perhaps it does signify that the JEI should avoid specializing in advocacy economics or apologetics of one type or another and concentrate upon positive description and interpretation if it is to remain a scholarly journal. That does not mean that the JEI should not publish quality pieces of evaluative analysis, only that it should be neither sectarian nor a journal of opinion. I am aware, of course, that all such lines are difficult to draw in practice. What is especially necessary is integrative work of both positive and normative types.

On Methodology

The JEI has published material widely differentiated not only in substantive content but also in methodological procedure. There are prose, mathematics, geometry, statistics, econometrics, comparative method, and even some poetry and humor (not all of which may have been so intended by the authors). Not all of these are appreciated by all readers; some have particular likes and dislikes in part governed by their form of alienation from orthodoxy and their own skills and interests. The journal, however, has been open and eclectic.

I think that the following should be said. First in my view, there is no permanent difference between orthodox and heterodox economics insofar as methodology is concerned. There are excesses (fudging) on both sides. The main differences are in the scope of admitted variables, or the length of causal-type chains, and in the combination of central problem and paradigm, especially central problem. Neoclassical orthodoxy is frequently criticized for its heroic abstractions, for example, its exclusion or neglect of certain institutional and behavioral variables. But the difference is not in the fact or degree of abstraction but in neoclassicism's focus on the market per se or on resource allocation through the market mechanism; institutionalist focus on the holistic and evolving system certainly deals with a different and larger problem, but its handling of *that* problem is no less and perhaps necessarily *more* abstract. At least that is how I often feel about my own work and how I assess the contents of the JEI.

Second, all research work involves some blend of induction and deduction, the differences arising in regard to the stage of their respective activity and/or form of their combination. Empirical work has been and indeed should be important in institutional economics. The landmark work of Wesley Mitchell, Morris Copeland, Gardiner Means, and John Blair, to name only a few, was primarily, albeit not exclusively, in this tradition. Continuous testing, statistical-econometric study, and the checking of analysis with the facts, such as they are (and not always quantitative), have an important place in institutionalist analysis. Permissive empiricism is no more justified in its work than in that of orthodoxy. Similarly, if it is nothing else, institutional economics is a body of theory and is often quite abstract.[20] Indeed, formalism and model building (not always so recognized) have been and are useful tools for institutional economists. They have their limits and are to be criticized when they are unduly confining. The point is that institutionalism necessarily has included and must include both empirical and theoretical work, but above all the concepts used must be adequate to reality and to the problem studied, questions on which reasonable (and unreasonable) scholars will differ. One person's reality is another's undue abstraction, and so on.

Third, it must be remembered that economic knowledge is at bottom probabilistic in character.

Fourth, not all the subject matter of institutional economics is amenable to quantitative empirical study, especially in a world of cumulative causation, but fifth, data-generating empirical work is important and desirable. Finally, the entry of subjective and metaphysical preconceptions is inevitable in all work.

On the Relation of Institutional to Neoclassical and Marxian Economics

It appears to be a presently unsettled question whether institutional economics, especially in its substantive investigations, is supplement or rival to neoclassical economics. The same question arises on the level of particular theories. Surely the different schools and their respective theories compete for attention and loyalty. The question of supplement vs. rival can be given no absolute answer; indeed, none is necessary. There is room for both views and for both schools: there appears no existentially necessary absolute incompatibility. They can at least co-exist. The JEI has been open to the work of holders of either view, which is almost totally irrelevant to manuscript evaluation.

The attitude of mainstream economists is not central to this report, although their goodwill is appreciated by all but the most alienated heterodox economist. Perhaps the ideal attitude is respect for substantive contributions rendered by heterodox economists and not condescension, token acknowledgement, rejection, or avoidance. But respect must be earned, although this does not mean that the contributions must be on orthodoxy's terms. Moreover, orthodoxy is by no means monolithic and closed.

My personal view is to dismiss the rejection of orthodox economics.[21] Neoclassical economics offers much constructive insight into the organization, operation, and development of the economy (market and non-market). A great deal of work by mainstream economists is relevant to the subject matter and interests of institutional economists. Orthodox economists have delved deeply into institutionalist topics. There are common albeit not universally recognized threads in neoclassical and institutional economics. Neoclassical economics may have considerable cultural specificity, but it does have trans- or inter-systemic relevance. The price system, however amenable to mechanistic and formalistic theorizing, *is* an important social institution, and its study *is* important. The great forte of neoclassical economics is to explain how markets *qua* markets operate in a pure or abstract sense. Given that exchange through markets is an important integrative system, that is no small accomplishment, and it is absurd for institutionalist-heterodox economists to dismiss their colleagues' work as irrelevant or trivial because some of their projects so seem. The operation of the market economy is a function of both market and institutional factors (including technology). Neoclassical theory is important for the understanding of the market mechanism, the operation of the firm, the role of so-called technical factors, and the allocative consequences of complex behavioral patterns. Constrained maximization analysis, which is not totally unrelated to pragmatism and instrumentalism, is useful in the study of institutional change. Government can be interpreted as an economic alternative and institutional change as a function of constrained-maximization calculations of advantage in power play. A wide range of neoclassical tools and analysis can be accepted and used[22] without limiting, presumptive optimality, or ideological factors, sometimes to produce quite unconventional results.[23] There can be, and indeed should be, an integration or synthesis of relevant orthodox and heterodox theories. Markets and culture are not mutually exclusive spheres. Logically contradictory explanations may contain factors which are in fact simultaneously present. Marketization in economic development is limited by, yet does tend to insinuate and transform, indigenous cultural and institutional relations and systems. There can be quite decent integrations of microeconomic demand-and-supply theory with analysis of the institutional factors structuring, channeling, and operating through markets, as is often done in the specialized fields, for example, the economics of medical care. The economic significance of rights is a function of both law and market. Institutionalist and orthodox theories of saving and capital formation do not have to be stated in mutually exclusive terms; they can be juxtaposed if not synthesized into something more heuristically and interpretively powerful. To acknowledge the foregoing is not, however, to say that the JEI should be an outlet for third-rate orthodox manuscripts. It does say that institutional economics can learn a great deal from neoclassical economics and that efforts at synthesis might be successful in advancing the discipline, if only by indicating the limits of rival approaches, although I think that the gain would be more than that.

Much the same can be said of the relation of institutional to Marxian-radical economics, the other principal occupant of the house of heterodoxy. Institutional economics shares many interests and subjects with Marxian-radical economics: evolution, holism, social relations, institutional change, conflict, power, and, inter alia, the factors and forces governing the structure of opportunity sets vis-à-vis neoclassicism's emphasis upon choice within extant opportunity sets.[24] That is the case even though Marxian and radical economists find institutionalists too liberal and neoclassically tainted to be radical.[25] Institutionalism has not been a road taken by disaffected younger economists. Marxian-radical economics has been more attractive; thus the Union for Radical Political Economics was formed. These economists were generally not students of institutionalists; often they had no or very little knowledge of institutionalism; and Marxism (in several variants) was available and seen as powerful. It seems that each institutionalist must do his or her own thing and find material and insight wherever located, whether in neoclassical or Marxian economics, notwithstanding the frequent neoclassical perception of institutionalism as radical (or irrelevant) and the Marxian perception of it as liberal apologetics.

All of what has been said above applies to positive institutional Economics; in the normative area, perhaps quite different considerations apply. What is important here, however, is that the JEI has been and should be open to substantive contributions of quality whatever their character, and in the quest for a substantive body of institutionalist analysis, heterodox eclecticism should not exclude orthodox and radical economics. *Economics* should include neoclassical, institutional and Marxian economics, with varying mixes of identification alienation, loyalty, and use. The JEI should retain its institutionalist-heterodox identity, but prospective contributors should not neglect these other schools of economics, including Keynesian and post-Keynesian macroeconomics (to which all that has been said above also applies), as sources of tools, knowledge, and insight. The pages of the JE1 should include a continuing dialogue, on open terms, between representatives of all schools – including those with a sense of belonging in more than one school – on subjects appropriate to its field of interest. Such a journal may be said to lack focus. In response to that view I urge that there is a wealth of substantive material already in the pages of the JE1 which (together with other material) warrants and indeed calls for integration and systematization. This view is supported by my own preliminary efforts to such ends. Moreover, with that in mind I am planning for the December 1977 symposium issue to be devoted to articles integrating and systematizing areas of institutionalist inquiry.

APPENDIX

This appendix attempts to formulate an outline or overview of the paradigmatic field of inquiry of institutional economics and its picture of the economy. The purpose is to indicate something of the enormous existing substantive content, and prospects, of a descriptive and interpretive institutional economics. In preparing this overview I have retained my strong desire to eschew any temptation to over-systematize and finalize institutionalism's or the journal's field of interest, as if that were possible, so as to avoid any substantive foreclosing of extant and future diversity. *Ex cathedra* definitions of a field of inquiry do not in fact determine what its practitioners actually do, and it is practice which should count, although attempts at definitional statements may capture descriptive accuracy of what has been done and/or help channel or focus future work.[26]

Institutional economics should be defined not solely by the orthodoxy whose critique it provides but also and primarily by its own conception of the economy, as its field of inquiry, upon which it renders both its critique of orthodoxy and its affirmative, substantive contributions. The emphasis in this appendix is on paradigm development and not its detailed elaboration into a corpus of substantive knowledge or hypotheses. Attention is directed to positive institutional economics, that is, to the subjects with respect to which institutionalist and other economists take normative positions, and to neither normative institutional economics nor the normative positions themselves. Thus, for example, it is necessary to differentiate the basic themes herein presented concerning the organization and control, and change of the presented concerning the organization and control, and change of economic system from specific positions thereon whether they be positive or normative. The emphasis is upon an outline of a positive paradigm,

The paradigm is considered to be complementary to that of neoclassical economics, say, in regard to the latter's treatment of resource allocation through the market. The paradigm is analytically rival to neoclassicism's with regard to institutionalism's emphasis upon the organization and control problem, except that: (1) the operation of the market is part of the organization and control field, as both a dependent and independent variable, such that neoclassical theory is, therefore, a contributor to the analysis of organization and control; and (2) neoclassicism, as a theory of choice, is carried further back to deep organization and control forces and choices.

For purposes of comparison, in abbreviated summary, the neoclassical paradigm centers upon the allocation of resources and distribution of income through the market: the operation of the market and market-like processes through constrained maximization and self-interested decision making typically involving voluntary exchange; methodological individualism; the conditions and processes of market equilibrium; and, inter alia, the conditions and processes of efficient (optimal) solutions. In contrast, the institutionalist paradigm centers upon a broader conception of the economic system than the market a wider and deeper range of variables insofar as its analysis deals with the problems of resource allocation and income distribution, and a different central problem, namely, the evolution of the organization and control of the (broadly defined) economic system. In some respects the institutionalist paradigm is capable of absorbing the lessons of neoclassicism; in other respects, it presents deeper and/or alternative explanations which can be seen as either complementary or rival. The paradigm may be outlined in the following way.[27]

A. The economy is an interacting subsystem of society which
   1. includes the market mechanism, where present, market-like processes, *and* the institutions which both form and operate through the market;
   2. is a decision making process governing the four basic economic problems of resource allocation; income distribution; level of income, output, employment, and prices (including the rate and direction of growth); *and* the organization and control of the economic system, or structure of power;
   3. is involved with the problem of order in society, namely, the continuing reconciliation of autonomy (freedom) and control, hierarchy and equality, and continuity and change, such that
   a. order is a process and not a condition
   b. order is a partial function of the market; *and* the market is a partial function of order;
   4. is a valuational process encompassing both the prices of goods (resource allocation) and the working rules governing access to and use of power (distribution of power) and thereby whose interests count
   a. civilization is a specific organization of values and of modes of valuational change;
   5. is a psycho-cultural system which includes
   a. power, knowledge, and psychology variables and their interactions
   b. such forces as technology, marketization and the pecuniary nexus, industrialization, institutional transformation, and the interactions of market institutions, and technology
   c. deliberative and non-deliberative forces and decision making
   d. methodological individualist and methodological collectivist forces and variables;
   6. in respect to all of the above is marked by deep general interdependence, that is, all variables and forces are both dependent and independent, through which systemic and evolutionary diversity originates and develops through cumulative causation.

B. The distinctive central problem of the paradigm is the organization and control of the total economic system, that is, the formation and reformation of the structure of power (and related knowledge and psychology variables)
   1. The economy is a decision making process with the ultimate continuing choice problem of its own organization and control
   a. System performance and policy issues are a function of both market and deep organization and control forces, processes, and variables
   2. The economy is a system of power going beyond the technical structure of industry (and industrial organization is more than a matter of price adjustments)
   a. The economy includes processes for the joint determination of both optimal solutions *and* the rights governing the substance of optimality, including constitutional and other rules
      i. Costs are a partial function of whose interests get registered as, and thereby become, costs to others
      ii. Market and non-market solutions are power-structure specific
      iii. The community's actual social welfare function is contingent upon preference and power-structure formation
         a. The economy is a game theoretic and not solely a maximizing system
      iv. The power structure governs whose interests count, say, in the market
         b. Power structure governs, and is governed by, the formation and distribution of productivity and/or welfare
   3. The organization and control of the economy includes facets of
   a. the working out of the meaning and reality of capitalism as a system, including its structure of power
   b. systemic and structural change
   c. social control and change beyond the market
   d. social control as power play

C. Insofar as institutional economics treats the same technical problems as neoclassical economics, for example, resource allocation, it stresses
   1. a wider and deeper range of variables[28]
   2. market operation and performance as a function of deep organization and control forces, rights, and institutions, including those ensconced within or forming the market itself
   3. the interrelations between market forces, individual choices, and institutions, each limited by, yet shaping, the others
   4. the endogenous determination of

    a. output definitions[29]

    b. preferences and values

    c. the formation of opportunity-set structures, vis-à-vis choice from within given opportunity sets, that is, the pattern of autonomy and of exposure to the choices of others.

In partial summary, the institutionalist paradigm focuses upon the deep problem of order and organization and control facets of the economy and upon an holistic and evolutionary view of the structure-behavior-performance of the economy understood to include all the above sets of variables in a system of general inter-dependence or cumulative causation. It thus focuses upon the methodological collectivist facets of the economic system, serving as a counterpoint to neo-classicism's methodological individualism but making use thereof as necessary,

    Two related areas of particular interest to institutionalist analysis are

D. Institutions, with respect to which study is undertaken of

    1. deliberative and non-deliberative processes of institutional change, including pressures upon institutions from within and without

    2. developmental and (comparative) performance theories and evidence

    3. institutional design strategy

    4. how institutions organize and channel production and exchange activity, with distributional and allocative consequences, for example, through their interaction with technology and markets

E. The economic role of government, including

    1. the legal-political system as a partial function of private economic activities, for example, the state as an economic alternative

    2. the deep legal foundations of the economic system, including the allocative and distributional effects of legal rules and rights

    3. the state as one facet of composite economic decision making, in part as a collective bargaining or exchange system and/or arena of power play.

The picture of the economy which is incorporated within or emerges from the institutionalist paradigm is that of a system of power, with elements of both conflict and harmony, and with conflict as both cause and consequence of economic evolution. It is, even more fundamentally, a picture of deep cumulative causation between market forces and institutions; of profound impacts of organization and control forces; of existential systemic diversity and open-endedness; of multiple social valuational processes, including the market; of inevitable and deep legal foundations; and of individual and collective action. The allocation of resources is seen to be a result of decisional forces and institutions which operate through and indeed form market supply and demand and which are subject to deliberate and non-deliberative reformation. It is a conception of the economy with respect to which the conditions of market equilibrium and the attainment of optimal solutions are but narrow, albeit important slices.

NOTES*

1. Corrected minutes of meeting of 19 November 1974; p. 2. It will be noted that the motion speaks of heterodox and not institutional economics. The idea of external journal reviews is presented in a letter by J. F. Bunnett, *Science* 189 (26 September, 1975). A suggestion to "analyze its contents and attempt to define its distinctive characteristics, to assess how far its product is differentiated from the output of other economics journals" was made by A. W. Coats (September, 1974, p. 599).

2. The report is in part based directly upon a complete (re-)reading of the past issues of the JEI except for the December symposium issues in which case detailed notes generally were used. Cognizance has been taken of materials to be published in the June, September, and December, 1976 issues. I have also benefited from conversation and correspondence with many persons concerning the tasks and accomplishments of the journal. The report itself was written during April, May, and June, 1976.

The personal character of the report, reflecting the vision of its author, must be acknowledged. Such an effort is a function of what one brings to it, expects to find, looks for, and sees. The heterogeneity of heterodox economics, to be emphasized below, underscores the importance of the role of the author's judgment. Such judgment is severely limited by, among other things, the author's lack of universal competence and expertise, including his certain inability to fully appreciate all that he has read in the past issues.

3. W. J. Samuels (December, 1975, p. 585).

4. The proceedings issue of March, 1972, technically edited at Michigan State University, was included in the first period tabulations, although the review article and notes and communications section were prepared under the present editor.

5. Forest G. Hill (June, 1967, p. 137).

6. Forest G. Hill (June, 1967, pp. 137–138). One article was on ideological responsibility in regard to the theory of capital and income distribution; one on wage-price guidelines; one on capital, savings, and economic development; another on development theory; two on the history of economic thought (dealing with John R. Commons and Wesley Mitchell); one on the anthropological study of economics; and one on the definition of leisure. There also were sixteen book reviews. (Book reviews were largely discontinued between the March, 1969 and March, 1972 issues, inclusive.)

7. Harvey H. Segal (September, 1969, p. 2).

8. Abraham Hirsch (June, 1967, pp. 79–82); Robert L. Heilbroner (December, 1970, pp. 18–20); and John S. Gambs (December, 1974, p. 956).

9. Edward Van Roy (December, 1974, p. 955).

10. David D. Martin (June, 1974, p. 272).

11. Article IX, section a: "The name of the organization shall not be used in any resolution or statement in connection with any partisan or political issue except insofar as said resolution or statement refers to a matter involving the purposes and objectives of the organization. Such resolutions shall require a two-thirds majority of the membership.

* All references not otherwise cited are to earlier issues of this journal, designated by month and year.

Such resolutions shall be so phrased as not to commit individual members." AFEE Constitution, as amended in 1970.

12. A. W. Coats (September, 1974, p. 604, n. 4).

13. Robert L. Heilbroner (March, 1975, p. 77).

14. John S. Gambs (March, 1968, p. 77).

15. For examples, Lauchlin Currie (June, 1969) and Abraham Hirsch (March, 1968).

16. W. J. Samuels, "The History of Economic Thought as Intellectual History," *History of Political Economy*, vol. 6 (Fall, 1974), pp. 305–323.

17. The foregoing paragraphs are based, in part, upon the following: C. E. Ayres (June, 1967); A. G. Gruchy (March, 1969); M. J. Ulmer (June, 1974); P. A. Klein (December, 1974); A. G. Gruchy (December, 1973, pp. 690–694); D. D. Martin (June, 1974, pp. 271–285); R. L. Heilbroner (December, 1970); S. Krupp (September, 1968); H. Wolozin (September, 1972, pp. 140–142); M. De Vroey (September, 1975, pp. 425–436); J. Dorfman (March, 1970, pp. 1–2; J. Jalladeau (March, 1975); and A. Randall (March, 1975, pp. 81–86). Also relevant are: W. S. Gramm (March, 1973), W. J. Samuels (June, 1975, p. 144); R. Lekachman (June, 1974, p. 267); W. Nutter (June, 1968, pp. 168, 169); A. Hirsch (June, 1976); W. H. Melody (June, 1974, pp. 287–290 and passim): R. F. Neill (June, 1969, p. 11); W. P. Strassmann (March, 1973, p. 167); K. E. Boulding (September, 1972, p. 127); W. Breit and K. G. Elzinga (December, 1974); D. Kanel (December, 1974, p. 920); V. G. Goldberg (June, 1975); J. J. Spengler (September, 1974); R. Solo (December, 1969); C. E. Harvey (September, 1972); W. A. Barnett (March, 1973); H. Sherman (June, 1975); A. Schweitzer (June, 1969); R. M. Solow (December, 1973, p. 695); W. Hochwald (June, 1973, p. 486); A. L. Dietz (September, 1975); R. A. Gonce (September, 1971, p. 88); F. Roosevelt (December, 1969, pp. 8, 10–18); E. J. Mishan (December, 1975, pp. 714–718); J. Barbash (September, 1967); K. de Schweinitz (December, 1974); G. Myrdal (December, 1974); and many others. See also W. J. Samuels (Ed.), *The Chicago School of Political Economy* (East Lansing: Division of Research, Graduate School of Business Administration, Michigan State University, 1976), chapters 16, 17 and passim.

18. Clifton, N.J.: Kelley, 1972; reviewed September, 1974.

19. See W. J. Samuels, "Ideology in Economics," in S. Weintraub (Ed.), *Trends in Modern Economic Thought* (Philadelphia: University of Pennsylvania Press, 1976).

20. D. Hamilton (June, 1973, p. 197); J. S. Gambs December, 1974, p. 956).

21. See A. Hirsch (March, 1968; June, 1976).

22. A. G. Gruchy (March, 1969, p. 7); K. de Schweinitz, December, 1974); S. M. Loescher (June, 1974, pp. 333–334); R. L. Heilbroner (June, 1974, pp. 251–252); P. A. Klein (June, 1973); G. C. Hufbauer (June, 1973); and R. E. Smith (June, 1974, pp. 419–420). See also A. E. Kahn (June, 1974, p. 307).

23. W. A. Barnett (March, 1973); W. J. Samuels (March, 1976, pp. 181–185).

24. R. S. Franklin and W. K. Tabb (March, 1974); J. Oser (March, 1974, pp. 167–170).

25. H. A. Sherman (June, 1975, pp. 247–248).

26. See A. W. Coats (September, 1974, p. 600).

27. The analysis uses a complex structure of primitive terms requiring specification, including institution, power, culture, market, and so on. See R. Brandis (September, 1975, p. 543).

28. That is, the total allocative system, the scope of the neoclassical model being a function of its limited definition of the discipline and not of the existential field.

29. For example, whether output includes worker sense of dignity and decision making participation.

INSTITUTIONAL ECONOMICS:
RETROSPECT AND PROSPECT, 1968

Warren J. Samuels

The essay published below was written in 1968. It was, I believe, the last major piece written before I left the University of Miami for Michigan State University. As stated early in the paper, the objective was to both demonstrate and make sense of Institutional Economics as a body of knowledge, as distinct from (but not unrelated to) being a protest movement and an approach to problem solving.

A short version of the paper was presented at a conference on the history of economic thought held at Duke in early December 1968; it was clearly too long by far for both the new journal, *History of Political Economy* and the struggling *Journal of Economic Issues.* When I became editor of the *Journal of Economic Issues* in 1971 it never seriously entered my mind to publish it, in part because of my early policy not to publish my own work in the journal and in part because of its length. Although encouraged by others to publish the paper in JEI, I was concerned that as editor some people would take what I wrote on this subject to be an attempt at an *ex cathedra* pronouncement, whereas in fact I was seeking to promote diversity and eclecticism in the *Journal.*

The paper was circulated privately among a number of individuals – most were conference attendees or Institutionalists – and continued to do so for some time. Some of the responses (slightly edited for typos) from those sent copies of the paper are worth recording. There was common agreement that the paper was infelicitously written; disagreement over its substantive content and import turned on school-of-economic-thought lines, but even among institutionalists even the objective of the paper was criticized.

Documents on Modern History of Economic Thought, Volume 21-C, pages 191–250.
© **2003 Published by Elsevier Science Ltd.**
ISBN: 0-7623-0998-9

Joseph Dorfman's note, dated 8 July 1968, was short: "I like your Duke address, very much. The only question is whether the first few pages are not too polemical."

John S. Gambs wrote of the paper (August 1, 1968), "I am enthusiastic about it" and invited me to be a panelist on an Association for Evolutionary Economics' session responding to his 1967 presidential address.

Clarence Ayres, somewhat to my surprise – because of its focus on power and the problem of organization and control, and on neither the interplay of technology and institutions nor the nature of a reasonable society, both important interests of his – told me in conversation, around 1972 or thereabouts, that he liked it very much, precisely because of its focus on power. Earlier he had sent me two letters, from Cloudcroft, NM, where he resided during summer prior to retirement from the University of Texas. The first, dated 15 August, 1968, said as follows:

Dear Professor Samuels:

Thank you very much for allowing me to read your long paper on Institutionalism during the summer, when I could take plenty of time to reflect on your analysis. In some respects this is the most thorough survey of the literature anybody has made. You make Mitchell's traditional inclusion intelligible, as well as his limited contribution, and make Commons's fundamentally institutionalist character more intelligible than he did. You say that you are presenting a shorter version of the paper at Duke in December. I hope that an unabridged text will be made available also, as I should like it to be available for purchase by the members of my class. At all events may I retain the copy you have sent me?

In some respects your interpretation of institutionalism makes better sense than anyone else's – even those of major institutionalist figures. This is true especially of your identification of the major concern of institutionalism with the most general aspects of socio-economic organization and the most general significance of the most basic socio-economic institutions. Your paper is most successful in establishing that such is indeed true of all genuine institutionalists – such, for example, as Maurice Clark.

To my mind the paper is weakest in its treatment of "standard" theory. Seemingly it is very generous of a writer who is not an avowed institutionalist himself to declare that the institutionalists have produced a body of knowledge that deserves to be taken seriously, no less perhaps, than the body of knowledge (micro cum macro) which standard (or traditional) economics has produced. But is this possible? Surely these two bodies of knowledge contradict each other. Or at the very least, some of their exponents on each side do so.

Can the judgment that standard theory is a significant body of knowledge worthy of respect and preservation be reconciled with the judgments expressed in Veblen's critical essays? I would say not. For example, it seems to me that what Veblen was saying in the celebrated monocotyledonous paragraph was that the whole trumpery of marginalism is of no more significance to post-Darwinian economics than Linnaean botany was for post-Darwinian botany. As a body of knowledge it has the same value as Linnaean botany.

The theory of the market in all its variants assumed an institutional system of which private property is the dominant feature. Adam Smith stated, quite explicitly, that such is the case, and if later exponents of the obvious and simple system are less explicit the reason is not just a taste for the low-level intricacies of market theory.

The truth is that all the institutionalists have been rebels to some degree – the same to which they have been institutionalists. As is suggested by your analysis, this was true even of Mitchell (who after all was Veblen's student), and most of all of Veblen himself. Their rebellion against conventional M&M theory reflects their dissatisfaction with the institutional system of which that theory is a projection. Most emphatically, there is something more to this division than degree of generalization. You are quite right in pointing out that there is a difference of degree of generalization between these two traditions. But in emphasizing this difference, even in an effort to do justice to institutionalism there is a danger of underplaying this even more important issue.

Ayres appreciated that the paper was not written as a polemic and that it represented the position of someone who believed that Institutional and Neoclassical economics were more or less supplementary and not mutually exclusive. He did not fully appreciate my commitment to Institutionalism ("not an avowed institutionalist himself"); he did not yet know of Gambs's invitation and of my contribution to the Association for Evolutionary Economics' program in December 1968 ("On the Future of Institutional Economics," *Journal of Economic Issues*, vol. 3 (September, 1969), pp. 67–72). And either he did not appreciate that the focus of the paper was specifically limited to Institutionalism as a body of knowledge, to the exclusion of Institutionalism as a protest movement against the market economy (I would now say a capitalist economy) and market economics (ditto) and as an approach to problem solving, or he felt that the meaning of Institutionalism as a body of knowledge is inextricably tied up with criticism of Neoclassical economic theory – a position especially crucial to those who emphasize mutual exclusivity. Of course, the overriding issue had two parts: the quest for social space, or status, and the content thereof.

My response to Ayres (August 27, 1968) consisted of the following, very much along the lines of (what I much later would develop as) social constructivism, theories as tools (not necessarily definitions of reality or Truth), and methodological and theoretical pluralism:

With respect to your comments: you say, "Surely these two bodies of knowledge contradict each other. Or at the very least, some of their opponents on each side do so." My reaction is this: the two bodies of thought are competitive insofar as they attempt to answer the same questions; insofar as they are directed to different questions, which they primarily are, they are not rivals, except for energy and attention; insofar as they are competitive, e.g. attempting to answer the same questions, they are not "contradictory" in the sense that one must be "right" and the other "false," rather the different scope of relevance of each is a major source of difference, which is at the same time a source of complementarity. My view is that the two bodies have and can continue to co-exist as bodies of knowledge however different their substance and thrust. With respect to their respective ideological coefficients, I quite agree that the one is conservative and the other liberal (i.e. Aristotelian vs. Platonic with respect to the relation of knowledge and social policy); though I could perhaps differ in that both (not just orthodox theory) reflect facets of the status quo, one directly and the other dialectically. No wonder, then, that their exponents contradict each other. As I point out later in the paper, Institutionalism has been both a body of knowledge and a movement of protest and reform; and a problem-solving approach, too. In the paper I have only endeavored to establish its meaning as a body of knowledge. What is an Institutionalist and what his relation to orthodox theory should be, are other and broader questions, questions which I hope to examine in a paper to be prepared for another meeting this fall.

As for myself, I was Ed Witte's student at Wisconsin, trained by Commons' students at Wisconsin in the middle 1950s, i.e. Witte, Perlman, Glaeser, Groves, Morton. I consider myself an Institutionalist with an appreciation for orthodox theory and with an appreciation of the limits of both; i.e. as an Institutionalist sans the alienation and rebellion, though I feel I can appreciate both the alienation and the rebellion. The future of Institutionalism lies in a constructive rapprochement with orthodoxy – certainly there are a multiplicity of areas in which it is being seen by orthodox economists that other, non-orthodox considerations must be attended to. Economic development and welfare economics are just two of the areas. On the other hand, I do not want too much ecumenism in economics (or elsewhere either): Institutionalism must serve as a check upon orthodox theorists – and rapprochement would not interfere therewith. But Institutionalists must do more than criticize: they have a field to develop, one which they have not really begun to fully develop, and one which must be developed. To too large an extent, its development has recently been by the high priests of conservatism.

I had enclosed a copy of my article on Witte ("Edwin E. Witte's Concept of the Role of Government in the Economy," *Land Economics*, vol. 43 (May, 1967), pp. 131–147) and Ayres, in a letter dated 8 October, 1968), wrote of Witte that,

> My only association with him was in connection with the AEA session of 1956, but that was sufficient to make me wish that I had known him better. I now see how you were able to make better sense of the institutionalism of the "Wisconsin School" than anybody else (including Parsons) has managed to do.

Witte had organized the session on institutional economics in his role as president-elect of the AEA. It is remarkable – as Malcolm Rutherford has found and documents – how little communication and interrelations had transpired among leading institutionalists during both the interwar and post-war periods.

Seeking permission to reproduce my paper for his students (which I had already given in my earlier reply), Ayres remarked that my "tolerance of conventional theory (much greater than mine) might serve to open some minds which would have remained closed to me."

> I must stop. But I suppose one might say that traditional theory is concerned with the allocation of owned resources by the institutions of sale and purchase; whereas institutional theory is the theory of institutional adjustment to continuing and accelerating industrial revolution. They are contrariwise in the sense that preoccupation with either is bound to be to the detriment of cultivation of the other. Thus my feeling is that "training" of professional economists in price theory unfits professional economists for what should be their

most important function: understanding, criticizing, and variously guiding the evolutionary process. The amazing thing is that so many economists manifest as much sense as they do notwithstanding their stultifying apprenticeship!

And we are teachers!

On October 11, 1968, I replying substantively as follows (both letters also discussed his forthcoming full retirement):

> I quite concur that training in traditional theory does not well fit economists to study institutional matters and that these latter are so important. As I believe I wrote you earlier, traditional and institutional economics *are* competitive in that respect, i.e. for energy and attention as well as loyalty. And cultivation of one, in a land of scarcity, does mean lack of cultivation of the other; it is, after all, a matter of readjustment of intensive and extensive margins. Most important of all, though, is the priestly function of orthodox theorists and their attempt to channel or disengage the evolutionary process – but those attempts are part of it too!
>
> My warmest regards and best wishes – and a hearty thanks for having been part of my education. One of the first books which I read in the period right after receiving my doctorate in 1957 was your *Industrial Economy*. Needless to say, it was not the first or the last piece of yours that I read – but it was high on my list for reading after securing the time for reading.

On 31 October 1968, Ayres wrote me that he had had my paper mimeographed and that students were being charged sixty cents apiece. He sent me two copies and offered copies to my students at the same price. (In 1997, Dan Hamermesh kindly arranged for me to receive a copy of Ayres' (?) – unmarked – copy from the Ayres Papers at the University of Texas library.) After noting his forthcoming full retirement at the end of the 1968–1969 academic year, he wrote the following:

> I'm glad you found *The Industrial Economy* worth reading. As originally planned, that book was to have contained at least an equal amount of "standard" introductory economics which, though presented in an skeptical spirit, would qualify the book as an introductory text. But my colleague and friend, Everett Hale, didn't come through with his part, so Houghton Mifflin went ahead with my part by itself – and, I'm afraid, lost money on the deal.

Ben B. Seligman wrote (August 20, 1968) that my paper was "just fine," though "rather long."

Alfred Chalk wrote (October 22, 1968) saying,

> In my opinion you have done a excellent job with a frustrating interpretive problem – finding the common threads that run through the fabric of institutionalism is a difficult task indeed. Parenthetically, having worked under Ayres, I think I am in a position to fully appreciate the difficulty of the task which you undertook.

Allan G. Gruchy wrote (September 1, 1968) that he was "very favorably impressed by" the paper. "You are one of the few people who can get to the core of institutionalism. So many individuals claim to be institutionalists or to be interested in institutionalism but when all is said and done they do not seem to be able to catch the essentials of the subject."

William Jaffé had written a doctoral dissertation on Veblen; I had known but not studied under him for several years (his father and step-mother had been at my wedding; the wedding present was Harrod's biography of Keynes). Jaffé thought (November 1, 1968) the paper too long, poorly written, and should be "made more digestible" though saying, "It is probably my fault that I have had such difficulty digesting it. I belong to the old school." He also wrote the following, much of it comprising a remarkable if not intriguing sermon, or programme:

> I don't think I can make any comments of value because it is so long since I have consulted the literature on the subject. In fact, when, to my surprise, I found myself cited in your paper, I pulled out from my bookshelves my tattered copy of the old AEA *Proceedings* to see what I had said thirty-eight years ago and to renew acquaintance with a long past and nearly forgotten self. I was glad to find that I am not in disagreement with my *alter ego*. I am grateful to you for that experience. . . .
>
> What I should prefer to see emphasized is the need to make institutional considerations an integral part of what you call "market economics." This should be done systematically. I cannot conceive of any good reason for presenting "Institutionalism" separately, as an autonomous discipline within the realm of economics. As you yourself admit, all that results from the valiant efforts of the Institutionalists from 1890 on, is an inchoate mass of "knowledge." Nevertheless, I believe that something can be salvaged from this rag-bag, insofar as it contains raw materials that can be used in the process of integration I have in mind. Really the Institutionalists *avant la letter* (e.g. David Hume in his essay, "Of Interest," and especially Karl Marx in virtually all his economic writings) made a better job of it and pointed the way to how it should be done. The devotees of modern Institutionalism, on the other hand, chose a different course. Instead of doing an enlightening job of exposing the implicit institutional substratum of the purely analytical models of "market economics," instead of attempting to probe to what extent the forces and relationships considered in these models are or are not *in specie aeternitatis*, i.e. ineluctable whatever the institutional setting, instead of investigating the possibility of directing modifications of institutions with a view to achieving greater efficiency in operation or greater justice and equity (which are themselves institutional concepts not at all independent of inherited ideologies, or of the state of the industrial arts or of the social order – all in constant flux), these devotees of modern Institutionalism simply struck a hostile and defiant attitude vis-à-vis analytical economics. With all the fervor of disciples of a newly revealed religion, they swore to extirpate the abhorrent false doctrines of the past. Naturally enough, the high priests of the established cult of analytical economics looked upon the upstart Institutionalists with contempt and would have no truck with them or their works. After all, their analytical models had a semblance of coherence and system, which was wanting in Institutionalism. Reason, therefore, as well as resentment and a concern for their vested interests, lay behind their intolerance of anything smacking of Institutionalism.
>
> Would it not be just as well to abandon the word Institutionalism and get rid of all the hampering associations that hedge the term? After all, ways of thinking about economic phenomena are institutions and possess all their attributes. That being so, names turn out to be important.

I should like to see institutionalism and analytical economics wedded, not in holy matrimony, but in a sort of free scientific union from which silly bickering is banished. An example of such a wedding is seen in the recent work of Demsetz, who so brilliantly brought analytical considerations to bear on the understanding of the institution of property – not private property, but property in general, including communally owned property.

This leads me to a final remark on the substance of your paper. I don't think it does any good to set out programs as you do in your third part. We want the show. A program without a show is an empty thing.

Now I come to the most testy part of my comments, which only goes to prove that I am an incorrigible, senile "square." I am offended by your lingo. Your opening paragraph alone struck me as a monstrous example of the worst in economic or sociological writing. It appalled me because of the havoc it does to our most precious heritage, the beautiful English language. I feel it is our duty, as scholars and men of letters, to cherish that heritage and try to make it more lovely still. When I see it defiled by academic jargon, I cannot refrain from protesting, even at the risk of appearing discourteous.

I don't know whether you will forgive me for this unseemly outburst. While it will pain me if you do not, I shall try to seek consolation in the hope that after further reflection, you may see fit to follow in the footsteps of those economists like Adam Smith, Henry Sidgwick, A. C. Pigou, Irving Fisher and Oskar Lange who couched what they had to say in an English that is a pleasure to read for its rhythm as well as its clarity and precision.

Jaffé concluded his letter with a request: "Would you please remember me to your colleague, my former student and constant friend, Victor Smith." – apposite because of the several sets of notes taken by Smith in Jaffé's courses published in these archival volumes.

On November 4, 1968 I replied to Jaffé as follows:

Thank you very, very much for your efforts in responding to my paper.

Your reactions were certainly testy but never unseemly. I appreciate your frankness and your coming to the point.

Certainly your comments are well directed toward the literary deficiencies of the paper. They need to be corrected. Without appearing to provide excuse, I would say that during the time the paper was researched, thought about, and written, I was extremely anxious about the subject: the materials of institutionalism *are* so diffuse and so varied, and so many have denied that it is a meaningful body of knowledge, I was seriously checking my inquiries from day to day by invoking the query, *is* there such a thing as an institutional economics? Perhaps I was unduly severe with myself, but for that reason and also because of the great heterogeneity of the subject-matter, I found the effort to generalize very difficult.

You reacted that I would have institutionalism separate from market economics. Here we have a semantic difficulty perhaps. It is precisely my purpose in the paper to show that Economics has had two traditions definitely not autonomous. If we need terms, Economics is comprised of market economics – micro cum macro – and theory of economic policy (wherein institutionalism primarily lies, though not uniquely) – theory of economic organization. The two have been wedded, more common law than otherwise.

Compare, for example, your reaction with that of Clarence Ayres, who also read the paper: Ayres believes that the two traditions are contradictory and cannot be so wedded. You, on the other hand, read into the paper that I would have two separate disciplines within the realm of economics. My position is this: two subject-matters generally differentiated by their primary problems (micro cum macro vis-à-vis theory of economic organization) but interacting (micro cum macro implications for theory of economic organization; economic-organization implications for micro cum macro) – pretty much as you go on to describe on page two of your letter, though I would go beyond that.

With respect to the last section: the problem of the paper is to interpret the retrospect and prospect of institutionalism, which I proceed to do in terms of institutionalism as a body of knowledge. But the prospect and retrospect of institutionalism is not only that of a body of knowledge but as a reform movement and body of criticisms (or orthodox, market economics). I offer in the final section not a program but a definition of the future prospects in terms of past developments. After all, institutionalism means different things to different people – it does have those different thrusts – and it would be highly presumptuous of me to limit an interpretation of the retrospect and prospect of institutionalism to what is perhaps its weakest facet.

With respect to the first part, I think it is crucial, though it obviously needs improvement: for therein I try to establish that the relation of institutionalism to market theory (micro cum macro) is not one of different methodologies but one of difference in scope and (primarily) central problem, ergo institutionalism as one aspect of the second tradition of economics, to wit: theory of economic policy. What makes the first section so unwieldy is that I felt a need to attend to the various facets of institutionalism which are so important (or so unimportant) to many different people.

. . .

One other reader of the paper, I might add [Robert V. Eagley], complimented me for what he saw as a non-romantic approach to institutionalism. However much the compliment is earned and accurate, much of the difficulty I had with the paper stemmed from precisely my attempt to treat institutionalism abstractly and analytically, in the face of its vast variety and in face also of the view that it could not be done, i.e. that there was nothing "there."

In conclusion, I am struck by your statement, "Any comment I should care to make upon the substance seems to be already contained in your paper – and that, in my opinion, is the trouble with it. It contains too much, all jammed together, so that little or nothing stands out." Perhaps I have tried too much; perhaps, I should have not attended to some of the things which institutionalism has been involved with and so on. I prefer to think that on the level of generality at which the paper operates, it is not too much, but perhaps needs to be presented better. And you have certainly stressed that!

Again, many thanks for your reactions. I think that your view of the relations between institutional and market economics is not terribly different from mine, that is to say, that the difference is semantic and not one of what economists should be or do.

I have given your regards to Vic Smith, who returns them with much warm feeling. As for myself, I look forward to seeing you at Duke next month.

The coda, as it were, to the foregoing took place in 1980 when Jaffé became the first Distinguished Fellow of the History of Economics Society and I had the pleasure of introducing him.

Allan G. Gruchy had to wait a year or so to send me his reactions, being immersed in revising his *Contemporary Economic Thought: The Contributions of Neoinstitutional Economics* for Augustus Kelley. On November 30, 1969 he wrote the following:

I find myself in agreement with most of what you have to say about institutional economics. My main concern is with your views about the need for a "general model" or body of economic theory which you think should be the content of institutional economics. I am not clear as to how far you would be prepared to go in trying to develop the concept of a "general" model of the total economy. You appear to think that institutional economists should somehow move beyond their theory of industrial capitalism to a general model of the whole economic system. At one point you say that the institutional economists seem to want to get away from a lot of the special or historical features of industrial capitalism in order to come up with a highly generalized theory of capitalism which would then approach being your "general model." As you have put it, your position is that "It is necessary, and possible, to construct a general model equivalent, with respect to the basic problems of Institutional Economic analysis, to the general micro and macro models, and performing the same functions. . . ." (p. 13).

I must say that I am somewhat dubious about getting very far with "effective system-level generalizing" beyond where we have already gotten. I do not think that institutional economists will ever be able to mathematize or econometricize their general model of industrial capitalism or of advanced industrial economies. Since there is no mathematics of dynamic change and the concept of economic process incorporates dynamic change, I do not see how we can ever get beyond a non-mathematical model of the economic process which will never be comparable to the micro and macro models of conventional economics – comparable, that is to say, in terms of abstractions. At the best I think that we can develop only system models of a limited nature such as a model for democratic western industrialized economies, a model for advanced industrial communist economies, a model for newly developing economies and so forth. If one tries to generalize on a higher level, I feel that the required abstraction will nullify much of the empirical significance of the model – or its usefulness in explaining the nature of the real economic world.

With regard to your general approach to institutional economics I am wondering if you are paying sufficient attention to the philosophical differences between the institutionalists and conventional economists. When we say that the institutional economists want to enlarge the scope of economics, they want to do so because they feel that economic reality can be better understood or grasped when it is regarded as a process rather than a static balance. The institutionalists and conventional economists have different views of economic reality which stem from their different philosophical positions – the institutionalists being evolutionary pragmatists while the conventional economists are at bottom static idealists. There is thus a wide philosophical gulf between these two kinds of economists which shows up in different views of economic reality and of the nature and scope of economics. To be sure, all economists are interested in "economic theory" and in "economic policy" but what separates them is a philosophical gulf which leads to radically different economic interpretations and economic policy recommendations. I have a feeling that you may be a little too prone to put all economists for various reasons in the same basket when in reality they have much not in common. Maybe I have misinterpreted your "common ground" of Knight and Commons and of Galbraith and Friedman (p. 20). While it is true that they are all interested in economic theory and economic policy, what is really significant is how they differ about the nature of economic reality and desirable economic policies. When you emphasize the need for more generalizing on the part of institutional economists I feel that you may perhaps not be emphasizing what is unique about institutional economics.

I appreciate very much your doing some very hard thinking about institutional economics. Unfortunately many who claim to be institutional economists seem to do little in the way of constructive thinking about our field of interest. . . .

Gruchy thus raised problems of substance and methodology that also arose in my later work on and in Institutional economics. I responded to Gruchy as follows:

> I want to thank you for the most penetrating critique of my paper on institutionalism which I have received. I quite agree – indeed, I was sensitive to these points when I wrote the paper – that the general model which I have had in mind may be impossible and that if possible so abstract as to be empirically, operationally, and practically of little utility; and, further, that, there are, indeed, differences between types of economists which my basket-lumping does injustice to and which cannot or should not be glossed over.
>
> Nevertheless, I think the effort to generalize is worth doing: while the probability is low, the payoff – if realized – would be large, and the effort, while frustrating at times, is basically interesting. I am trying to work out a model of the economic system as a system of power (decision making; mutual coercion) which would *at least* identify and generally (that damned word again!) relate the critical variables and fundamental choices involved in the economic organization of society.

The project to which I refer led to the publication of "Welfare Economics, Power and Property," in G. Wunderich and W. L. Gibson, Jr., eds., *Perspectives of Property*. University Park: Institute for Research on Land and Water Resources, Pennsylvania State University, 1972, pp. 61–148.

Jack Barbash, in addition to saying that the paper was too general (I somewhat disagree) and wordy (I agree; due in part from applying certain fundamental themes to different but related topics), had some interesting substantive criticisms: That the paper proves commonality of interests, not of substantive application to diverse situations with similar results (I agree) – as, he thought, was done by Neoclassical economists (I disagree). That the paper did not offer enough of what Institutionalists had to say (I agree but wonder whether it could or should have). That Neoclassical Economics was a body of logic applicable to prediction, and that Institutionalists wanted a viable body of thought with which to provide dissent from and a gloss on Neoclassical Economics, whose theory was too much driven by consistency and other objectives (by which I think he meant what I have since come to call the Neoclassical research protocol of generating unique determinate optimal equilibrium solutions). Barbash's last point illustrates the inexorable, and not altogether undesirable but potentially overdone, hold which protest and criticism has had on Institutionalists. Whereas my objective was to neither deny nor denigrate Neoclassicism but to focus on the positive: Institutional Economics as a body of knowledge with its own basic problem, that of organization and control, or power. Be all that as it may, the construction given in the paper to Institutional Economics as a body of knowledge has served to no small degree, but by no means solely, as the basis of both my lectures on Institutional Economics and many of my later writings. Anyone familiar with my own subsequent work on both the economic role of government and topics of Institutionalist theory will readily perceive both their prefiguring by and consistency with Institutionalist themes presented below.

I should again acknowledge that I wrote the paper with some anxiety. This was due in part to my recognition that I could not deal with everything important to all Institutionalists, and that some people might resent omission of their approach or interests, and in part to some angst over the fact that I was dealing with fundamental questions on which more established and famous people had written extensively. (Similar concerns had affected the writing of *The Classical Theory of Economic Policy* several years earlier.) These considerations led to the numerous early qualifying statements.

It should be clear to any reader that the essay attempts only to identify a domain identified in terms of a central problem, and is not comprised of a particular theory of any aspect thereof. In that sense, the essay is paradigmatic and programmatic. It is to be expected that multiple theories and models on each point could, and should, be developed; also that different paradigmatic statements of the central problem could be articulated. It should also be clear that the analysis is predicated on the view that in some respects Institutional and Neoclassical Economics are supplementary and that in some respects they are competitive.

Among other things, I would now emphasize the following. First, Neoclassical Economics has its own conceptualization of the problem of organization and control, and that while that conceptualization constitutes in one sense its deepest meaning, its preoccupation is with the allocation of resources through the price mechanism of pure conceptualized markets and with the generation of unique determinate optimal equilibrium solutions. Second, the methodological differences between the two schools are more complex and subtle than is examined here. Third, the degree of sophisticated methodological understanding among general economists (not specializing in methodology) was, alas, overstated in the paper. Fourth, the degree of respectable diversity within both economics as a whole and Institutional economics itself, especially in the former, has shrunken in the years since the paper was written. Fifth, some attention has recently been given to aspects of the problem of organization and control to so-called New Institutional Economists, though generally, to varying degrees, still within the Neoclassical paradigm and its attendant ideology. The work of Ronald Coase and Douglass North in particular are unusually pregnant, and certain to have a place in any future serious, generalized theory of organization and control: North's attention to power and ideology in the formation and evolution of institutions, and Coase's emphasis that institutions matter and that markets and costs, so far from being given, are a function of numerous social and legal variables, not least the decisions made by firms over the division of labor between firms and the markets within which they operate (Coase's work is somewhat paralleled by that of Gardiner C. Means). All these (and still other themes) are critical to a more complete corpus of organization and control theory.

The paper, published here for the first time, has not been substantively revised; only minor stylist alterations and corrections have been made. The original referencing system has been retained. I am indebted to Margaret Henderson for computer scanning the original manuscript.

INSTITUTIONAL ECONOMICS: RETROSPECT AND PROSPECT

Warren J. Samuels

1. Historic Meaning of Institutionalism
    a. protest movement re:
        i. market economy
        ii. market economics
    b. problem-solving
    c. body of knowledge
2. Institutional Economics as a Body of Knowledge
    a. relation to market economics
        i. methodology
        ii. scope:
            a. scope of variables
                i. different or broader answer to same problems
            b. difference of central problem studied
                i. thesis of two traditions of economics
                    a. micro cum macro
                    b. theory of economic policy
    b. Institutional Economics as a Theory of Economic Policy
        i. basic problem of organization of the economy as a system
            a. market as institutional complex and operating within and in inter-action with other institutional complexes
            b. basic economic problems' resolution a function of institutions arrangements and not market forces alone
                i. institutions as regulatory systems
            c. institutional organization of economy as fourth basic economic problem
                i. mutual impact of the organization problem upon the others and of the others upon it
                ii. organizational problem one of distribution of power in society

203

    a. importance of structure of power
    b. part of larger problem of order
    c. holistic or system-level, and evolutionary, view of economic process
    d. power structure as phenomenon of collective action; embodiment of problems of freedom and control, and continuity and change
  iii. psychology and knowledge as correlative to power as main facets of Institutional Economics
  iv. further consideration of the problem of power
    a. functioning of the working rules
    b. interrelation of legal and economic processes
    c. value theory of Institutional Economics:
      i. relative rights governing power structure (allocation of power)
      ii. working rules governing both relative rights and transmutation of private into social interests and vice versa

The objective of this paper is to establish the accurate retrospective and feasible prospective meaning and identity of Institutional Economics. Notwithstanding the not inconsiderable professional knowledge about Veblen, Mitchell, and, to some extent, Commons, the meaning and thrust of their work as Institutionalists has remained obscure and enigmatic. Part of the difficulty lies in a general tendency to explicate the meaning of Institutionalism in terms of the specific theories formulated and developed by particular Institutionalist writers when the theories are either ambiguous, conflicting, and/or non-comparable. This paper will attempt, contrariwise, to establish the meaning of Institutional Economics in terms of the basic problems with which it has dealt, with a view to the prospective specification of a general corpus of Institutionalist theory – more precisely, a general model – concerning these problems. More generally, this paper will attempt to articulate the meaning of Institutional Economics as a body of knowledge, and thereby show the possibility of a general and comprehensible model of Institutional analysis as well as its significance.

I

An assumption of this paper is that there is a coherent and viable body of thought which may be designated as Institutional Economical. If this view is to prevail, and the objective of the paper be realized, the author will have to overcome the widely held view that Institutionalism is so diffuse and ambiguous as to be ethereal and phantasmagoric.

It must be granted that the major individual Institutionalists are each difficult to interpret as well as difficult to compare and integrate. The thrust of each is multiple and their respective styles are quite different. Institutionalism meant different things to each of them; indeed, no one of them was either apparently certain in his own mind or unambiguous in his writings as to its meaning. Even the concept "institution" was differently and ambiguously defined. In addition, the several strands of Institutionalism are colored by the personalities of the dominant figure in each as well as by the particular subjects with which each chose to work and by the research procedures deemed appropriate and/or necessary in exploring these subjects. Against such an amorphous existence it is no wonder that several commentators have concluded that Institutionalism is no longer a distinct, viable, and going concern as an organon of thought and inquiry, and, indeed, that it may never have been. Thus, in the early 1930s Paul Homan delivered his well-known dictum "that an institutional economics, differentiated from other economics by discoverable criteria, is largely an intellectual fiction, substantially devoid of content"[1] and Addison Cutler, in the late 1930s, although defending Institutionalism against the false conclusion that it did not exist, improperly drawn from the admitted difficulty of definition, nevertheless concluded that Institutionalism was dissolving and ebbing.[2] More recently, Kenneth Boulding has treated Institutionalism as largely passe, as a movement of dissent with small direct impact albeit very great indirect influence but as something to be spoken of mainly in the past tense;[3] and Lafayette Harter has concluded that Institutionalism is moribund.[4]

Further reason for the diffuse and consequently ambiguous identity of Institutional Economics lies in its heterogeneous character and checkered reception. Institutionalism has been a corpus of knowledge, as this paper shall try to make clear; but it also has been much more. Although not equally so, Institutionalists generally have been critics of and reformist towards both the market economy (or finance capitalism) and market economics.[5] For some, like Edwin Witte[6] and the Commons tradition generally, it has represented a problem-solving approach to questions of public policy; but for others it has represented an avenue of dissent, "an economics, nay, a whole social science, of rebellion against established habits of thought."[7] As Gambs has put the matter, the Veblenian tradition in particular has attracted "inconsequential rebels, non-conformists, reformers, and dissenters;"[8] and Dowd, both "the loosest kind of crackpots . . . and the most circumspect of pure researchers."[9] For some it has represented escape from a theory to which, for various reasons, hostility had developed; for others, a secure dissent as a badge of alienation and rejection;

for others, a mode of more meaningful and constructive analysis of market structure and operation, and, jointly or severally, study of cyclical conditions; and for still others, a criticism and rejection of ideology masquerading as knowledge. In general, it has reflected a greater consciousness of the relative ubiquity of policy choices in human affairs coupled with a Platonist view of the relation of knowledge and social policy but coupled also at times with pessimism and at other times with optimism. But if Institutionalism has often had a posture of astringent and derisive hostility and contempt for market economics, mixed with social reformism and a quest for knowledge, and ergo has been heterogeneous; market economics also has been similarly heterogeneous as well as unreceptive and at times directly antagonistic toward Institutionalism. Market economists have been exclusivist, haughty, smug, dogmatic, and condescending toward "institutional" studies, which have been sensed as a threat to the identity of market economics; have taken refuge in "pure" models, in a sense at times no less withdrawn than the most despairing Veblenian; have been something of a high priesthood of a conservative-individualist apologetic; and have had a basic general (predominantly micro) theory nonetheless decidedly heterogeneous in the variety and diversity of formulation of central theorems and propositions in almost every area thereof. Yet Institutionalists have been haughty and condescending in their own way; too often toward an abstract theory which they were unable or unwilling to master, work with, and amend. The Institutionalists have yielded much that was cavalier, pejorative, and destructive; no small proportion of the rejection of Institutional Economics has been due to its often negative, non-constructive or uncooperative, and superficial criticism.[10] But if there is an "irritating quality of institutionalism," the "source of irritation," as William Jaffé wrote long ago, has resided within market economics itself, in "any attempt to explain economic phenomena in ways other than those consecrated by the customs, habits, and usages of the guild of economists. One is almost tempted to regard opposition to institutionalism as institutional in nature."[11] Gentility, humility, and intellectual pluralism often have been lacking on both sides; but one must say that they also have been present on both sides. In any event, the heterogeneous quality of Institutional Economics as a social phenomenon has contributed to its lack of intellectual identity. If *rapprochement* can be established upon the basis of a definite subject matter, so much the better for the profession.

Institutional Economics thus has represented and signified reform and dissent; and has manifest the ambiguity-creating ambivalences characteristic of its multifarious origins and existence. But Institutional Economics has also encompassed a body of knowledge, and while that body of knowledge remains inchoate, it promises a lasting and fundamental contribution to the discipline and to the social sciences generally. Not that it will have provided final and conclusive answers; rather that it will have helped at least to formulate problems, the problems in terms of which Institutional Economics as a body of knowledge has meaning.

Let it be granted that much of what unequivocally are distinctive Institutionalist writings is essentially reformist, it nevertheless remains valid that Institutional Economics contains a corpus of knowledge in the same sense(s) that market economics contains a corpus of knowledge. (On the question of methodology, see below.) Considered as reformist writings, the Institutionalist literature may be identified as Platonic[12] with respect to the relation between knowledge and social policy, that is to say, concerned with creating ideals which may serve as criteria for altering the status quo reality. Just as Platonism is idealism, or knowledge of what might or *should* be, Aristotelianism is knowledge of what is, that is to say, knowledge of extant reality. The distinction is not absolute: idealist knowledge is derived from or upon the basis of status quo reality and bears the imprint thereof;[13] and realist knowledge has idealist elements contained in it through the valuational elements in theory construction and testing, including the choice of problem and the erection of ideal-types.[14] Insofar, then, as Institutional Economists produced *is* knowledge, such knowledge is generally on an epistemological par with that of market economics; and insofar as Institutional Economists produced reformist or Platonic or idealist *ought* knowledge, such knowledge may have fundamental realist or *is* knowledge distilled from it. In both cases, the Institutionalists may be said to have dealt with a particular set of meaningful problems. It is in terms of that set of problems that the meaning and identity of Institutional Economics will be formulated in this paper.

It will be contended herein, then, that there is a distinctive set of problems, or area of knowledge, relating to Institutional Economics. Although it will be more thoroughly developed later in this section, it should be made clear at this point that the subjects comprising the meaning of Institutional Economics as a body of knowledge have been the object of attention and analysis by writers neither commonly nor even remotely considered Institutionalists. That is to say, while there is a distinctive subject-matter to Institutional Economics, non-Institutionalists have also dealt with that subject-matter, including writers generally considered to be market economists. As will be suggested later, the field of economics historically has been comprised of (for present purposes) two major strands, one clearly formulated and recognized and the other not: the former, micro *cum* macro, and the latter, the theory of economic policy – with numerous connecting links. Institutional Economists were not alone in contributing to the latter field; nor, of course, did they fail to contribute to the former complex of fields.

The author is thus arguing both that Institutional Economics has substance transcending the specific theories of particular writers generally acknowledged to be Institutionalists, and that writers other than such acknowledged Institutionalists have made contributions to the subject-matter involved. It may be anticipated that some of those who consider themselves Institutionalists and some of those who consider themselves market economists (and yet fall into the category of having made contributions to the subject-matter with which Institutional Economics deals) will not be entirely pleased with such contentions. Institutional Economics will be envisioned as having lost whatever identity it has to the professed Institutionalist; and the identity of economics will appear bifurcated to the market economist. But it is absolutely essential if the meaning and identity of Institutional Economics as a body of knowledge is to be established and, more important to the discipline, if its subject-matter is to be nourished and more adequately developed in economic analysis generally.

Several problems impinge upon the construction of this paper and should be specified.[15] First, the author is sensitive to and believes that he fully appreciates the fact that Institutional Economics, to many, is more than a body of knowledge (just as to many others it is not appreciably if at all a body of knowledge). The author does not presume nor intend to disparage Institutional Economics as a practical-problems approach or as a critical and reformist movement. His intention is simply limited to identifying the meaning of Institutional Economics as a body of knowledge.

Second, since it is the objective of the author to interpret the meaning of Institutional Economics as a body of knowledge in terms of certain basic problems, not only should the specification of those problem in terms of the content of particular Institutionalists' theories be avoided but also their definition in terms of the present author's research interests and conceptualizations, the latter of which is a particularly pervasive and ineluctable problem since any interpreter is limited, in an ultimate sense, to his own vision. Still, the effort has been attempted.

Third, since the meaning of Institutional Economics will be rendered in terms of certain basic problems – or definitional scope, as will be seen shortly –, those problems should not be over-systematized. Rather, they should be precise enough to be meaningful as a definition of scope but open enough to allow for theorizing with respect to them, as well as for their own definition and redefinition, to be worked out. This paper, then, is *not* an attempt to suggest or construct a closed system.

Fourth and finally, it must be remembered that the subject matter of Institutional Economics, as will be defined in Part II, below, is perhaps infinitely more expansive and complex than micro and macro theory, and that, consequently, there is a tendency – reflecting a set of real needs, to be sure – for analysis to progressively extend into and become both a theory of society and a theory of history. "Big questions" have that tendency.[16] As Frank Knight has pointed out, the subject matter "raises vast and difficult problems," as it "must deal with behavior forms and social processes that are much less tractable intellectually than are market data or even utility comparisons."[17] As Commons (among the other Institutionalists) and Knight have pointed out, the subject matter is further complicated (e.g. for predictive purposes)[18] because of the diverse, ramified, and unintentional consequences of human action (particularly collective action) and also because of the open-ended character of volition or the will-in-action. As Knight put it, "To have a mind means to change it occasionally; hence to act unpredictably – but not too often, too erratically, or too far, or it would cease to be mind. As intelligent beings, we live somewhere between causation and chaos."[19] It is, after all, "man himself"[20] who allocates resources, distributes income, makes for one or another level of income and rate of growth, and, of profound significance for present purposes, molds his institutions. That the subject matter of Institutional Economics has been and remains, therefore, difficult is easily ascertainable, if only by its market lag behind the systematization ant comprehension that market economics has achieved; but that does not mean it is impossible – only that it takes somewhat longer to work out.

The author would further contend that the *essential* and *permanent* distinction between Institutional and market economics is *not* a question of methodology.[21] It certainly must be acknowledged that historical disputes over method have been frequent and heated. But it can now be said that these disputes primarily represent spillovers from: (a) arguments over an axiomatic and presumptive laissez-faire posture of conservative market economists; (b) continuing and productive methodological or epistemological explorations of significance far beyond the Institutional vs. market economics controversy and affecting both indiscriminately; and (c) the actual fundamental distinction between Institutionalism and market economics, namely, the question of scope (see below).

Dispute over empiricism (including quantification) and verification is no longer a matter of serious importance. Not only is it recognized that quantification where feasible is salutary, but it is also recognized both that not all quantitative analysis is Institutional and that not all Institutional analysis is quantitative.[22] Similarly, it is now almost universally recognized that induction and deduction are mutually necessary and complementary; and that verifiability is a *desideratum*. Furthermore, arguments over the "abstract" character of market economics are (or should be) now recognized as largely: (a) beside the point (Institutional Economics *is* abstract also; see below); and (b) spillover from controversy over scope. Thus, the question of greater "realism" – aside from the significant question of the scope of the economically relevant – Foes not appear to remain in this respect (i.e. abstractness) a serious and differentiating methodological variable.[23]

Methodological disputation involving Institutionalists has also included arguments concerning the appropriateness of physicalist or mechanistic analogies and equilibrium models as working assumptions or tools. Here the actual issue is not epistemological but substantive: it is not a matter of the methodological credentials of knowledge but of the explanatory power and accuracy as well as heuristic value of the asserted knowledge.[24] The same is true, inter alia, of the competition and rationality assumptions.

This is not to deny or negate the important methodological contributions of Institutionalists in the past. Institutional Economics, by its methodological criticisms, has helped foster corrections in the methodology of market economics, including greater quantification[25] and verification requirements and less presumptuousness concerning valuational (ideological) and/or inaccurate substantive assumptions, analogies, and nuances. In these and related respects, Institutionalists have helped perform a vital service for the discipline. (And, of course, it should go without saying that Institutional Economic analysis is not quite fully guaranteed "against the tyranny of any categories of thought.")[26]

By the present day, it seems to be clearly and generally recognized that complementariness and ineluctable composition characterize methodological positions and procedures,[27] and that any juxtaposition between Institutional and market economics does not permanently rest on absolute methodological differences. Both Institutional and market economics represent blends of induction and deduction, of variable admixtures of intensive and extensive margins of abstraction ant concreteness or generality ant specificity.

What is necessary and, moreover, possible though presently lacking, is a coherent general body of theory delineating Institutional analysis and serving as a vehicle and framework for research, interstitial exploration, specification, and theorizing and speculation generally. As the difference between Institutional and market economics is not methodological, it can also no longer be doubted that the literature of Institutional Economics contains important bodies of theory. The abstract and theoretical character of Institutional economic analysis – that it is not simple or pure empiricism – is rather widely recognized. With respect to Commons, although his theoretical writings are difficult, perhaps often "obscure and cumbersome" and a "tangled jungle of profound insights,"[28] his major works are theoretical and abstract, granted a function of "observation and experience, not . . . introspection and a priori surmise,"[29] indeed, "one of the most abstract theoretical systems in American social thought."[30] Theoretical also is the Institutional work of C. E. Ayres and J. M. Clark, and of Veblen as well. Although there is opinion to the contrary,[31] and although, like Pareto,[32] Veblen was very much of an armchair theorist, his was, after all, as Heilbroner insists,[33] a *theory* of the leisure class. So also was Selig Perlman's a *theory* of the labor movement. Institutional economics has been theoretical and has the intellectual meat of a group of theories.[34] Edwin Witte's belief that Institutional Economics was more a method of approaching policy questions than a body of theory did not mean that it did not have reasoned and developed theories.[35]

But as Witte[36] and many others have pointed out for many decades, there is no coherent general corpus of Institutional analysis, no general model within which particular theories have interpretive significance and which gives analytic meaning to the field as a body of knowledge and area of theorizing and research. It is necessary, and possible, to construct a general model equivalent, with respect to the basic problems of Institutional Economic analysis, to the general micro and macro models, and performing the same functions of enabling the working out of fundamental concepts of problems, enabling the analysis of relationships, providing a conceptual framework for particular studies, and enabling integration and tentative synthesis. While such a defining and integrating framework would channel theorizing and research efforts, it would promote rather than shackle analysis, as in other areas where paradigmatic models function as continuing albeit typically incrementally revised integrating frameworks.[37]

There has been much lament by both Institutionalists and non-Institutionalists over the absence of such a general heuristic model together with considerable support for its development.[38] But none has been forthcoming or, at least, none has been forthcoming that has been accepted; and although several writers have maintained that it exists in embryonic form,[39] enthusiasm for its development has been mixed with uncertainty and pessimism as to the likelihood of its realization.[40] It would appear that the importance of the subject matter with which Institutional Economics has dealt, developed in Part II, *infra,* is ample justification for effort to create a viable and coherent body of theory. The present author would differentiate between constructing the general model of theory and defining the meaning of the field in terms of basic problems, though the difference is subtle and realization of the latter would be a first step in effectuating the former. It is the hope of the author that the present paper may contribute to the formulation of a general body of Institutional theory[41] by helping to identify the meaning of Institutional Economics in terms of the basic problems comprising the subject matter of the corpus of Institutional theory.

The difference, then, between the mainstream of market economics and what has been called Institutional Economics is not a question of methodology – at least as a matter of fundamental and thereby permanent controversy – but rather one of *scope* and, what is closely related, of the central problem(s) of inquiry. The distinctive contribution of orthodox or traditional economics has been its analysis of how *market* forces[42] function to continually resolve the basic economic problems of resource allocation, income distribution, and the level of income (including such related considerations as rate of growth, price level, and employment), together with consideration of relative-price, relative income, and income-level and related consequences of government fiscal and monetary policies whether or not deliberately compensatory or remedial; hence, *market economics*. Now, Institutional Economists have been interested in the same basic economic problems,[43] though to some extent they have developed somewhat different answers, the differences in same cases being great and in others small;[44] quite aside from the substantial variance of answers among market economists themselves, each answer more or less within and consistent with the general micro and/or macro models as they have developed. So far as the Institutionalists are concerned, analysis of non-competitive conditions and economic instability are cases in point. Part of the differences in the Institutionalist answers are consequent to the different (viz, wider) scope of the variables encompassed in their analysis, and part are no doubt due to differences in analysis per se (as in the case of market economists). The differences notwithstanding, the Institutionalists have more or less readily acknowledged the importance of the basic economic problems and the need to explain their modes, etcetera, of resolution in a market economy through market forces. Here the primary (but not exclusive) ground of disagreement has been over the descriptive accuracy (realism in one sense) of the assumptions of the traditional model as the latter developed, e.g. the descriptive quality of the competitive assumption. Part of the variance and contrariety in result is also the consequence of models dealing with variables allegedly neglected by (though not outside the scope of) market economists, e.g. the relation of technology to saving and capital formation.[45] So that Institutionalists' analyses are competitive with those of market economists in the sense that they attempt to come to grips with or answer the same problems; but the Institutionalists tend to have a different (wider) definition of relevant social space and thus a wider range of variables. In these cases generally, as with cases in which the Institutionalist has simply postulated different relationships between the same variables, the question involved is one of relative explanatory power and descriptive accuracy. In all these cases the theories of the Institutionalist and market economists are competitive in that they generally attempt answers to the same questions.

But the difference which has been fundamental, in the view of the present author, is not just that the Institutionalists have come up with different (including broader) answers to the *same* questions. Rather, the distinctive, viable, and coherent quality of Institutional Economics is that it has or may be said to have focused on *different*[46] problems and questions. Whereas market economists have been primarily (but not exclusively – see below) concerned with market resolution of the three basic economic problems of resource allocation, income distribution, and aggregate-income determination, the Institutional Economists have been primarily interested in not only how the market resolves the basic economic problems, but how other institutions participate in the process of resolution and, above all, in the view of the present author, in a fourth basic economic problem, which may be alternately called the problem of organization and control, the problem of order, or, inter alia, the problem of the distribution of power.[47] As will be further developed in the following section of this paper, the Institutionalists have interpreted the economy as including more than the market as defined by market economists; have interpreted the meaning of economy as more than the conventionally recognized basic economic problems; have interpreted the resolution of the conventionally recognized basic economic problems as a function of variables larger in number than those considered under the aegis of the market; have thus interpreted the meaningfulness of market economics as part of a larger corpus of economics; and finally have interpreted the larger and transcendent fourth basic economic problem as that of the changing organization of the economic system as a whole.

The difference, then, between Institutional and market economics has been the combined one of both scope and difference in central problems studied; ultimately, that the Institutionalists concentrated upon the basic problem of the organization of the economy as a system including the market and the impact of the resolution of *that* basic economic problem upon the three conventionally recognized basic economic problems and of the latter three upon the former, hence the greater scope.[48] Now, what *is* of profound and permanent significance is that market economists have also considered questions concerning the organization of the economic system as a whole, or, as sometimes expressed, the institutional framework with and within which market forces operate and interact. Indeed, the market-plus-framework postulate has been fundamental to market economics. *But* the crucial point is that market economists have not included consideration of the fundamental,[49] fourth economic problem (and therefore also the greater scope of relevant variables) in their general body of theory. That body of theory primarily concerns market resolution of the three conventional basic economic problems and *not* the organization of the economic system as a whole either as a separate problem or in terms of its impact on market resolution and the impact of market resolution upon organization. No general model or body of theory concerning the latter has been developed, except generally and incompletely with respect to, first, the market *character* of the western economy, and, second, doctrinal expositions concerning the role of government "in" a market economy.

That market economists, however, have considered non-market phenomena is of profound and permanent significance. Consideration of the institutional framework, of *system* phenomena, has always been part of economics although hitherto it has not been formally considered in economic models,[50] due largely to the micro-macro scope thereof; i.e. exogenous to economics *qua* "theory." There is a vast literature but no general theory or model. The subject-matter of Institutional Economics thus is not unique to Institutionalists: while Institutionalists have concentrated upon it, non-lnstitutionalist writers have also devoted attention to it. Institutional Economics thereby encompasses more than the Institutionalists. (See note 52.) *Neither* group has developed a general corpus of theory or a general model of the fourth basic economic problem. The permanent and fundamental significance of this is that economics, which started out (upon its distillation from moral philosophy or ethics) as political economy, has always had not one but *two* traditions, that of the theory(ies) of market forces, encompassing micro and macro, and that of the economic system as a whole and its organization and control mechanisms (including the market). This coexistence of two traditions has been obscured for many reasons, not the least of which was ideological; but also the fact that the system of western industrial capitalism was the object of study meant that the larger study could be obscured by the smaller, which eclipse also tended to result from the fact that market economies developed a general body of theory concerning market forces and neither market economists nor Institutional economists developed such a body of theory in the larger area. In any event, economics has not been only the science of the operation of the market, though it has been most conspicuously (and productively) that. It has also been, haltingly, incompletely, unsystematically, and (nevertheless) dogmatically at times the study of the system as a whole, of which the market has been but one part.

If "economic theory" – meaning micro *cum* macro theory, both broadly defined so as to include capital theory, monetary theory, shifting and incidence theory, distribution theory, and so on – has been the obtrusive mainstream, then, the present author would argue, the "theory of economic policy"[51] has been the unobtrusive but omnipresent second subject-matter of economics, secondary but present in the works of market economists, though not as the central problem, and primary in the studies of Institutionalists, largely as the central problem.[52] The second tradition, the subject matter of Institutional Economics – theory of economic policy, is reflected, for example, in the recent literature on the theory of economic policy of the classical economists;[53] and is manifest in the writings of such economists as Frank Knight, Ludwig von Mises, F. A. Hayek, and Milton Friedman, all of whom, no less than Commons and Ayres, albeit to different doctrinal and policy conclusions, have presented analyses of the economic decision making process considering the market as part of a larger economic system; and is manifest as well, e.g. in the literature of welfare economics generally, including the study of the institutional mechanisms which transmute private into social values, costs, and benefits.[54] That is to say, works of Knight et al. deal with the subject matter of Institutional Economics as interpreted in the following section of this paper, as do works of Commons et al. In the view of this writer, the concept "theory of economic policy," fully specified, is practically co-extensive with the concept "Institutional Economics,"[55] however non-systematically both have been developed by market and Institutionalist economists. The subject matter of both comprise the second, undernourished but profoundly important tradition of economics. The juxtaposition of this second tradition to market economics suggests if not fully demonstrates that the economics of the market is the economics of a subsystem[56] and that an economics of the system as a whole – obviously substantially different from, but not exclusive of and unintegrated with, micro and macro – not only has always existed in generalized and relatively inchoate form but is a common ground of economists (like Knight and Commons, or Galbraith and Friedman) who are otherwise ostensibly worlds apart. These economists have their positive and nonnative differences and disagreements; but what they fundamentally have in common will be interpreted below as the basic problems composing the meaning of Institutional Economics, an Institutional Economics which is, first, clearly economics, and, second, comprised of economist contributors beyond the Institutionalists as historically understood. It would appear, then, that the Institutionalists have been specialists in an area in which market economists have treaded (and treaded relatively extensively though almost totally eclipsed by the arena of their main concern) but, alas, have been no more successful in developing it as a coherent corpus of knowledge or general model. The following section will attempt to specify the identity of Institutional Economics as a body of knowledge in terms of a distinctive subject matter comprised as a set of basic problems, the problems of any theory of economic policy but as distilled from the writings and theories of the Institutionalists.

II

Market economics, as the study of market resolution of the conventionally recognized basic economic problems, has made several fundamental contributions to economics, including Institutional Economics. Those contributions center upon, and are embodied in, the functional definition of the economy in terms of those basic economic problems; the concept that the economy is a decision making, or choosing, process with respect thereto; and the ubiquity of incurrence of opportunity costs along multifarious intensive and extensive margins. This insight is grasped by Institutional Economists and is extended to apply to a much more extensive and, indeed, pervasive set of phenomena. As already indicated, the Institutionalists interpret the economy as encompassing more than the market, and interpret the meaning of economy in terms of considerations broader than the conventionally recognized basic economic problems. The Institutionalists thus recognize the importance of individual (and subgroup) valuation and preference, and the operation of the market mechanism. Their point of departure, however, is that the "market system" – meaning the historical western economy – may have two different contextual connotations. First, it may mean simply the market mechanism, either as it exists in the real world or as it is hypothesized or idealized in market theory; and second, it may mean, more complexly, both the market mechanism and the institutional apparatus which implements the market mechanism and which impinges upon the mechanism and, thereby, upon individual (and subgroup) preference both per se and as it is manifest in the operation of the market mechanism. Since the Institutionalist is primarily concerned with: (a) the market; (b) the institutional apparatus of the economy; and (c) their total and changing configuration of interaction, the Institutionalist is also interested in the fourth basic economic problem, the continuing problem of structuring and restructuring the economic system. Thus, the Institutionalist defines the economy more broadly so as to include the fourth basic economic problem; extends the scope of social policy or choice to include the very institutional structure of the economy itself; and acknowledges opportunity costs not just in terms of resource allocation or factor-of-production use but also in terms of relative power and the moral, customary, and legal rules which govern behavior and relative power.

Comprehension of this economic-system approach of the Institutionalists requires or may be facilitated by an understanding of the Institutionalist view that the market itself is a complex of institutions and, further, that the market as one complex of institutions is literally one among several major clusters and, still further, fundamentally interacts with them. Before examining these themes, however, the meaning of the concept "institution" must be provisionally clarified. This is no easy task, of which the history of Institutional Economics and commentary thereon is evidence. Consider, however, several of the more well-known attempts at definition or characterization: "the more important among the widely prevalent, highly standardized social habits;"[57] "widely prevalent habit of thought;"[58] "group habits of thought;"[59] "socially prevalent habits which in any given group standardize the behavior of individual members;"[60] "pattern of inherited habits;"[61] "collective action in control, liberation, and expansion of individual action;[62] "*methods of action* arrived at by habituation and convention and generally agreed upon;[63] "group of customs that are related by serving some particular interest;"[64] "habitual modes of activity and conduct [which] have grown up and have by convention settled into a fabric of institutions;"[65] "modes of association in which men live together in society;"[66] and, inter alia, "accustomed ways of doing and thinking, `habits of action and thought widely current in a social group'."[67] These meanings all involve, each in its own way, many difficulties of specification and differentiation, and of inclusion and exclusion. There is no need in this paper for anything like a final definition of the concept; rather, the concept, like that of "value" in market economics, will have its definition worked out as the body of theorizing develops. But what these attempts at definition and/or characterization seem to involve at bottom, it is suggested, is this: human arrangements in terms of group thought and behavior patterns which, formulated and articulated in custom and in law, govern both the choosing by individuals and the interrelations ultimately between individuals. Institutions, in other words, govern or influence both demand curves and cost conditions directly, and thus the resolution of the conventionally recognized basic economic problems; and also the organization of behavior or structure of power, and thus the fourth basic economic problem and the other three indirectly as well. The Institutionalist juxtaposes to the spontaneous economic actor of market theory the variable but at any time socially given modes of behavior and organization, both formal and informal. Institutions are more than "habits of thought;" as Veblen himself recognized, they include "methods of action." Commons' definition, of "collective action in control, liberation, and expansion of individual action," is not fully satisfactory or devoid of difficulties, but it is widely encompassing and does pregnantly and seminally point to the problem of the distribution of power: habits of thought and methods of action not only influence demand and supply but they also channel behavior and regulate interpersonal relations, which relations ultimately involve questions of relative freedom and relative power. Finally, *inter alia*, the definitions and characterizations more or less nicely epitomized in Commons' definition point to the problem of order – the reconciliation of freedom and control, and continuity and change – which is the larger meaning of the problem of the distribution of power, and of which more later. Indeed, the meaning of "institution" is umbilically related to those larger problems which are at the core of the subject-matter of Institutional Economics. In "concentrating" upon "institutions" the Institutionalists treated the fundamental organization of society and economy as their central problem. The organization of economy is embodied in "collective action" in all its forms.

What the Institutionalists argued, then, was that the market itself was an institution – an organized way of doing things, of thinking, and of structuring relations –, or, more completely, that the market itself was a complex of institutions. These institutions are for the most part rather commonplace: they include private property, contract, the corporation and other forms of business organization, labor unions, negotiable instruments and credit generally, and, inter alia, the labor market and the wage system.[68] So commonplace they are taken for granted in the modern equivalent of the natural order; but they are important: they have a history and that history is not only one of conflict but specifically one of jockeying for power in a changing, evolving, man-made organization of society and economy and structure of power which evolves in part because of that very jockeying for position. But it is not simply power that characterizes these institutions; it is also that of which power is itself a manifestation: the grand problem of social order. In either respect, these institutions together comprise the larger institution of the market, and as the former change so does the latter.

Technology – an institution, the application of knowledge for practical purposes – has been a force for the integrative cementing of social relations but it has also been a disruptive and divisive force; it is a major source of change. The corporation, so clearly an institution involved in structuring the market, and so clearly also a primarily market phenomenon, is also "a social and political institution whose influence extends far beyond the market it serves."[69] And money, the pecuniary dimension of the market (and of any other modern) economy, is to Gambs the very "nucleus of institutional theory."[70] Mitchell was the leading Institutionalist theoretician of money (as he was also, in his studies which are not distinctively Institutionalist, a leading market theoretician of money).[71] To him, "Orthodox economics is a `logic of the institution of money economy masquerading as an account of human behavior'."[72]

> It will become evident [Mitchell wrote] that orthodox economic theory, particularly in the most clarified recent types, is not so much an account of how men do behave as an account of how they would behave if they followed out in practice the logic of the money economy.[73]

The purpose of the author at this point is not to inject Institutionalist criticism of the psychological assumptions of market economics; rather it is to point to

Mitchell's declaration of the far reaching consequences of the market institution of the money economy. He thus followed the sentence quoted immediately above with the following:

> Now the money economy, seen from the new viewpoint, is in fact one of the most potent institutions in our whole culture. In sober truth it stamps its pattern upon wayward human nature, makes us all react in standard ways to the standard stimuli it offers, and affects our very ideals of what is good, beautiful, and true."

Indeed, he went on,

> The strongest testimony to the power and pervasiveness of this institution in molding human behavior is that a type of economic theory that implicitly assumed men to be perfectly disciplined children of the money economy could pass for several generations as a social science.[74]

What makes the market economy what it is, at any point in time and over time, are the institutions which give it substance, the institutions within which and under whose aegis market forces germinate and operate. The market is an institutional phenomenon, a complex of subprocesses themselves institutional in character.

The complex of market institutions – the complex of institutions which yields the market – is but one of several such complexes. Needless to say, the others include organized religion, the vast areas of custom both sanctified and unsanctified by moral rules and law, and the legal process itself. Particularly, it is the legal process which came under the investigative eye of Institutionalists from Richard T. Ely to Walton H. Hamilton, from Robert L. Hale to John R. Commons, and from John M. Clark to Edwin E. Witte. Indeed, it is the paramount importance of law and the political process as a whole from which it emerges as a cultural phenomenon which made "political economy" so attractive a nomenclature for several Institutionalists.

This view of the legal process should not obscure the correlative and in some ways even more paramount importance attributed to custom. It has occasionally been alleged that the Institutionalists grossly neglected the role of custom, in preference to the field of law.[75] No one, however, can be familiar with the work of Veblen, or the ideas of Commons – to whom custom was a major source of law, particularly as the legal process had to choose between conflicting customs (including between old and new customs), and to whom custom and sovereignty were the two correlative forms and sources of working rules – and fail to recognize that custom was given a substantial place in the Institutionalist general model of society and economy. The misapprehension, the author would suggest, is a result of the recognition (and, of course, activist policy posture and participation) by Institutionalists that, particularly in an age whose spirit was one of change but *in* any age, institutions, whether or not they themselves originated in deliberate design, would come under scrutiny and be subject to alteration by changes wrought deliberatively and by design (i.e. that the "natural history" of institutions was a product of both intended and unintended consequences) and this increasingly through the action of the state. Institutions, which always have been and always will remain a blend of deliberative and non-deliberative elements, were seen as becoming increasingly subject to a more profound policy consciousness, expressed in part in a volitional psychology, but ultimately in

> ... the emergence of the concept of good or bad political economy out of mythical entities such as nature's harmony, natural law, natural order, natural rights, divine providence, oversoul, invisible hand, social will, social-labor power, social value, tendency towards equilibrium of forces, and the like, into its proper place as the good or bad, right or wrong, wise or unwise proportioning by man himself of those human faculties and natural resources which are limited in supply and complementary to each other.[76]

In particular, it is the opportunity costs and the underlying necessity of choice involved in the very structuring of custom and of law and of law in relation to custom that is a major though subtle thrust of Institutional analysis, for both Veblen and Commons, and all others. But the increasing marginal significance attributed to the legal process, recognized by the Institutionalists as always important but accelerating over the last several centuries, should not eclipse the acknowledged powerful role of custom in economic affairs: it governs much of the stuff of life. The complexes of institutions embodied in what has here been called the legal process and custom functioned reciprocally and correlatively as parts of the economic system as a whole.[77]

The important points that the market is a set of institutions and that this set of institutions co-exists with still other sets of institutions thus leads to a third important point, namely, that there is pervasive and inevitable interaction and mutual dependency existing between these several institutional complexes. Though Veblen at the margin concentrated his analysis on the interrelation of the forces and institutions of the market and custom, and Commons, on the interrelations between market and legal processes, both found that the exclusion from formal market theory (though not, at least in retrospect, from the theory of economic policy of market economists – though most Institutionalists would no doubt insist that the latter remained naive) of such interrelations as variables rather emasculated the scope of relevance of the conclusions of market theory. But the fact of differential intensive and extensive margins is not at issue; it is, rather, a strength to the profession. What is important, and needless to insist or dwell upon at length, is that one of the major thrusts of Institutional analysis has been that the market does interact with other institutional complexes which have direct and fundamental economic significance, particularly with respect to the structure of economic power, i.e. the organization and control of economic activity. If market forces germinate within institutions, so too do institutions germinate from market behavior in the historic Western economy. The evolution of the rent bargain, the price bargain, and the wage bargain – fundamental to the devolution of modern industrial capitalism and to the modern economy generally – was shown by Commons to have resulted from the interaction of market, customary, and legal forces, ergo, at the margin, the legal foundations of capitalism.[78]

One set of problems with which Institutional Economics has been concerned as a body of knowledge, then, has involved the evolution and operation of the institutions which make the market what it is and in terms of whose change the market undergoes transformation; and the other institutional complexes with which the market interacts, which interaction also effectuates changes in the participating institutions, both market and non-market. The Institutionalist thus has been concerned with problems which center on the institutional organization of capitalism – on capitalism as an institutional system – and on the interaction and interrelation of capitalist (or market) and other institutions; or, to use a widely employed dichotomy – to whose artificial separation the Institutionalist often objects –, the Institutionalist studies the market and the framework (both as institutional complexes) and their interaction and interrelation as parts of the total economic decision making process; or, still further, the Institutionalist studies both the private and the collective aspects of economic behavior.

What this means, to carry the analysis a step forward, is embodied in a major theorem of Institutional Economics, a theorem which, perhaps more than any other in the field, dramatically characterizes part of the substance of Institutional Economics as a body of knowledge. Since Institutional Economics encompasses the study of market, of framework, and their interrelation and interaction with respect to the resolution of the basic economic problems, the relevant theorem of Institutional Economics is that the resolution of the basic economic problems is a function not only of the operation of the market – and perhaps not even primarily, albeit immediately or proximately – but also of the institutional framework. The proposition that the basic economic problems of resource allocation, income distribution, and aggregate-income determination are resolved in part through the operation of the institutional organization of economy per se, is a major heuristic principle of Institutional analysis. Although his statement equals some by market economists in point of exclusiveness, Ayres has given a forthright and clear articulation of the issue and relative position of the Institutionalist:

... the object of dissent is the conception of the market as the guiding mechanism of the
economy or, more broadly, the conception of the economy as organized and guided by
the market. It simply is not true that scarce resources are allocated among alternative uses
by the market. The real determinant of whatever allocation occurs in any society is the
organizational structure of that society – in short, its institutions. At most, the market
only gives effect to prevailing institutions. By focusing attention on the market mechanism,
economists have ignored the real allocational mechanism.[79]

If one will discount the position taken therein on the metaphysical question of
the relative importance of market and institutions as allocational mechanisms
– or resort to some kind of Marshallian-type model in which both have their
place as do demand and supply factors in the Marshallian short and long run
–, the central proposition is clear and unequivocal: resource allocation, etcetera,
is a function of both market forces and the institutions which comprise the
market and those which interact with market institutions and forces. As
expressed elsewhere by Ayres, "our economy is not organized by the market
mechanism. The market is organized by the economy. The order which the
market exhibits is derived from the organizational patterns of the economy and
is an expression of such order as actually obtains in the economy."[80]

This fundamental view, for present purposes a statement of a central problem
with which Institutional Economics is concerned as a body of knowledge,
pervades the Institutionalist literature: the resolution of the basic economic
problems is a function of the institutional framework directly and also indirectly
insofar as the structure of the market itself (as a decision making or power
process) is a function of the institutional organization of the economy. The
central proposition is reflected elsewhere, for example, in Loucks' "generalization that, although the operation of basic economic principles transcends all
types of economic organization, the specific forms that their operation takes
on and the specific social consequences which flow from their operation
differ widely from one type of economy to another,"[81] a proposition which is
substantially equivalent to that of Schumpeter, in which economic laws are said
to "work out differently in different institutional conditions, and that neglect of
this fact has been responsible for many an aberration,"[82] which is quoted by
Gordon illustrating Schumpeter's interest, parallel to that of the Institutionalists,
in "the interaction between economic behavior and the evolving institutional
environment."[83] Thus Dorfman writes of Commons' *Distribution of Wealth*
as "emphasizing customs, the role of fixed social relations and legal rights as
basic factors controlling the operation of the marginal principles . . ."[84] and
Chamberlain, of how in Commons' theory of collective action, "Exchanges of
goods and services must take place, on terms reflecting the relative propertied
advantages and disadvantages of the negotiators . . . within a framework of

collective law and custom, so that collective action has in fact structured the relationship . . . [such that] The distribution of economic goods depends on the way in which collective action affects these three types of transactions [bargaining, rationing, and managerial], which collectively exhaust all kinds of economic activity."[85] What was to the classical economists a struggle over the corn laws – "The practical problem was whether the power of the state should be used to maintain the incomes of the farmers and landlords, or whether the import duties should be reduced to safeguard the incomes of manufacturers and merchants"[86] – has been to the Institutionalist the generalized problem of income distribution as always a function of relative opportunities which are a partial function of relative rights which in turn are a partial function of legal and other institutions. Just as Copeland has pointed to the institutional performance or rendering of the central economic management functions,[87] Hale has argued that income distribution is a function of relative coercive power which in turn is a function of relative legal status, that is to say, that distribution is a function of institutionally produced or supported coercion even in a supposedly non-coercive state.[88] Resource allocation is a function of market forces but market forces reflect the "conditions of social relations among the owners and exchangers of commodities and services."[89] "Institutions," wrote Gambs,

> in the Veblenian system, tend to give a semi-permanent coercive advantage to certain groups at the expense of others. . . .. As any culture changes, the advantages of one group may count for little in a new environment, while formerly submerged groups may rise to the top of the scale of coercive power.
>
> The nature of coercive power differs from institutional system to institutional system, and probably varies in complexity directly with the complexity of the culture. The fact that persistent coercive advantage is useful in the acquisition of goods and services, endows the advantage itself with something akin to economic value, and sets men who do not possess the advantage to thinking how they can acquire it – or negative the advantage possessed by others, or acquire a superior advantage. In short, men strive to create new rights, or to alter existing institutions or legislation because of the economic advantage involved.
>
> An awareness of the capacity of institutions to change is an awareness that the rules of the economic game may be changed. This circumstance may often alter the ground of economic conflict, from a conflict over things to a conflict over proposed or presently existing arrangements.[90]

It is this dependence of the resolution of the conventionally recognized basic economic problems upon the institutional organization of economy which accounts for the attention given by the Institutionalists to the fourth basic economic problem, the distribution of power. For the operation of market forces itself is a reflection of the distribution of power and, what is more, this means that the operation of the economy is deeply involved in the resolution of the problem of order, or, to put the matter differently but to the same effect, the resolution of the problem of order has profound and ubiquitous impact upon the operation of the economy whether defined narrowly to include just market forces or defined broadly to encompass also the institutional organization of the market and the other institutions with which the former interact.

Deferring for a moment the problem of order in relation to Institutional Economics, it is now possible to specify several levels of Institutional analysis. Quite obviously, Institutionalism encompasses studies of particular institutions: their genesis, development, transformation, and functioning. On the next level of greater generality are studies concerning either the relation of particular institutions to the system as a whole or the interrelations between different institutional complexes. Finally, the level of greatest generality encompasses studies of the system as a whole with more or less distinct properties or characteristics as an economic system. Institutionalism includes all three types of studies; for present purposes this taxonomy should suffice.

Herein lies the general thrust of Institutional analysis: since the Institutionalist contemplates the economy as encompassing more than the market, Institutional Economics has as its ultimate frame of reference the organization of the economy as a whole, including all the economically relevant institutions and their interaction – most distinctively the institutions concerned in one way or another with the distribution of power both within the market and between the market and other institutions.

This emphasis upon the economy as a whole, as a system more encompassing than the market of market theory, has been underscored by Gruchy's adoption of the descriptive phrase, "holistic economics." That term, he wrote in what has become a classic inquiry into Institutional Economics as a corpus of thought, "calls attention to what is more characteristic of the new economics: its interest in studying the economic system as an evolving, unified whole or synthesis, in the light of which the system's parts take on their full meaning."[91] Institutionalism thus studies "the economic system as a whole rather than as a collection of many unrelated parts,"[92] as a "total cultural complex."[93] (By looking at the economy as a whole, and thereby as a cultural phenomenon, the Institutionalist has been enabled to throw in relief such distinctions as between knowledge and cere-monialism; between technology and institutions, about which see the concluding section; and between making goods and making money; the latter of each of which pairs is seen as fundamentally reflective of the particular culture or institutional organization of economy, and the former as something more transcendental or less culturally relative.) Thus, Mitchell had written that "all studies of special institutions become organic parts of a single whole;"[94] and Ayres, that "all the institutions of any given society form a continuous functional whole."[95] Although there is opinion to the effect that Institutional Economics is not holistic,[96] and although Institutionalists are somewhat open to Boulding's charge of premature synthesis,[97] and although, too, there has been little effective system-level generalizing, it would seem that there do co-exist both holistic or system-oriented studies and studies of particular institutions.[98] What is distinctive of Institutionalist works on particular institutions is the tendency to refer to the larger system (just as there is in market theory a tendency of particular studies to relate to general equilibrium models) and, what is to some extent the same point differently put but very important nevertheless, to get into questions of the distribution of power.

Holism thus signifies the greater scope of Institutional analysis; with the economy as more extensive than the market, to include nonmarket institutions of economic relevance and to treat the economy as an institutional system. Although Commons' three later works (*Legal Foundations of Capitalism*, and particularly *Institutional Economics* and *Economics of Collective Action*) were something of an attempt at system-level integration and model construction, no Institutionalist has produced a permanently satisfying model or one that has "caught on." (Reference may be made to the works of such writers as Talcott Parsons,[99] Alfred Kuhn,[100] and Robert A. Solo[101] as exemplifying system-level studies in the Institutionalist sense.) Absence of total or permanent success notwithstanding, the Institutionalists' emphasis upon resource allocation, etcetera, being a function of both market and non-market institutions underscores the holism of their analysis. (This abstract holism or system orientation also further evidences the theoretical character of Institutional Economics.) As a body of knowledge, Institutional Economics has been concerned with capitalism as a system not just with its most distinctive feature, the market as conceived by market economics. Thus, Gruchy elsewhere wrote that "The correct definition of institutionalism is that this type of economics is a study of the disposal of scarce means within the framework of our developing economic system. It rounds out economic science by providing a theory of the going economic system, or, in other words, a theory of capitalism. Institutionalists study institutions only as subsidiary parts of a larger matrix in the form of the economic order . . . what is of prime importance to the institutionalist . . . [is] a theory of capitalism . . . institutionalism is primarily a positive, creative movement which aims at broadening the nature and scope of economic science by pushing beyond basic theory to create a theory of our developing economic system."[102]

The view that the economy is more comprehensive than the market is aptly expressed by Seligman's proposition that, "It was the institutionalist contention that the market was not the sole area for economic action."[103] Once it is understood that the Institutionalists were attempting an holistic model, or at least had an holistic working conception, of the economy, it becomes readily apparent why the Institutionalists found that the hallmark of that larger arena of economic action was power. The reason lies in the utter necessity of considering the problem of order – of which the problem of the distribution or structure of power is one facet – once one no longer takes as given either the institutional organization of the market or the institutional complexes with which market forces and institutions interact. Consideration of institutions inevitably must involve consideration of what it is that institutions do and consideration also of change either in the substance or ends of what institutions do or of the structure of institutional organization itself. Since – aside from the impact of institutions on such things as taste and thereby on demand functions – the main relevant function of institutions concerns their involvement in the distribution of power (both within the market and between the market and other institutions), the Institutionalists almost (if not absolutely) invariably dealt with the struggle for power and the jockeying for position over changing the status quo distribution of power as part of the subject-matter of economic knowledge. What are for most people ideological truths were for the Institutionalists manifestations of the underlying, pervasive, and perennial struggle for power (that is to say, rationalizations of positions in that struggle), but not something devoid of social significance because of their ideological character. They are indicators of a basic social problem and its attendant processes rather than positions to be taken for granted or as immutable.

In his characterization of Institutionalism, Jaffé long ago came very close to the problem of power as a facet of the larger problem of order. Wrote Jaffé:

> Only when an author uses institutions not as so much constant background but as dramatis personae in the economic play of forces and only when he portrays changes in the character of these dramatis personae as capable of profoundly affecting economic relationships, do we find institutionalism properly speaking.[104]

The key expression is "economic relationships," for it is precisely the non-price and non-quantity relationships which are abstracted from by market economics that Institutionalism concentrates upon; and, furthermore, the problem of order in Western society is very largely the problem of reconciling stresses and strains of "economic relationships" considered in a context of power.

One of the truly seminal treatments of the problem of order in the literature of the profession is that by Joseph Spengler, according to whom the problem of order involves the reconciliation of the "somewhat incompatible" combination of autonomous participation, necessary coordination, and necessary continuity,[105] which the present author has restated as the reconciliation of the dual basic social problems of freedom and control, and continuity and change.[106] As Spengler points out, the problem of order is not solved in any manner either pre-ordained or exogenous to man's behavior but rather through the coordinating role of working rules, value-attitudes, and institutions generally.[107] The deepest problem, therefore, of Institutional Economics, as with any theory of economic policy, has to do with the impact of the processes of resolution of the problem of order upon and in economic affairs, with that part of the problem of order which has to do with economic affairs and with economic affairs as they involve the problem of order. At its core is the problem of power, i.e. of the structure of control in economic organization and of control over the choice of change, ultimately including change of the distribution of power itself. Since institutions are involved in the structuring of power, and since institutions perform the coordinating function producing order, the subject-matter of Institutional Economics thus embodies as fundamental considerations as may be found in the social sciences. As a body of (Aristotelian) knowledge, Institutional Economics does not simply take the existing order of social affairs for granted but rather inquires into the processes who produce that order, including the economy.

It is with respect to the problem of order, therefore, that the foremost premise of Institutional Economics is the necessity of a theory of collective action. Paraphrasing John Maurice Clark, control is an integral part of economy, without which it could not be economy at all. The one implies the other, and the two have grown together.[108] The problem is not freedom vs. control but the structure of power, or of the pattern of freedom and control, or of the pattern of freedom and of exposures to the freedom of others, and the further problem of continuity vs. change of that structure and of those patterns. These problems are resolved through the continual achievement of order, and the mode of resolution is collective action in control, liberation, and expansion of individual action, to again use Commons' phrase.

Institutional Economics thus deals with *conflict* and with *order*, ultimately with the very foundations of social life. Since it must consider the problems of freedom and control, and continuity and change, it must have elements of a theory of social control and a theory of social change, in both of which, and in Institutional Economics generally, conflict issues will be central and prominent. As a theory of social control, Institutional Economics has had to consider the genesis and operation of working rules – of custom, of morality, and of law – and the small and large conflicts and bargains which are wrought out over conflicts in those rules. Seen as a system of power, the economic process is examined in part as a struggle for power; as a system of power as a check on power; as a structure of power exercised in part through the forces of the market and in part through institutions, market and other; as maneuvers to capture and repel institutions active in the arena; and, inter alia, as a power structure with differential power holdings and with differential access to social control institutions. Seen as a system of ubiquitous mutual coercion, the economic process is examined in terms of the foundations and practice of coercive power; of coercive power and opportunity differentially structured by the working rules which are themselves influenced by mutual coercion as an object thereof. Yet institutions are not only instruments of mutual manipulation; they perform organizational, civilizational, and conflict-resolving functions. There is a dynamic interaction between behavior and institutions, a major source of pressure for change in the process of balancing continuity and change. Expressed differently, the market mechanism (e.g. under competition) was seen as but one of several regulatory systems in society. More globally, power play and social control are not independent processes.

Now, the purpose here is not to recite an Institutional theory of the problem of order, or of freedom and control, continuity and change, power, coercion, or social conflict. The purpose is rather to establish these problems as the central distinctive subject-matter of Institutional Economics – and any theory of economic policy – as a body of knowledge. As Spengler shows, the problem of order has not been left unattended by the major historic schools of economic thought. But the Institutionalists, notwithstanding their failure to produce a coherent general theory or model, are distinguished by their concentration upon this set of problems. It is in formulating particular theories on these subjects that the Institutionalists were often ambiguous, incomplete, and in disagreement. But, it is a major theme of this paper that these subjects were the crucial part of Institutional Economics and were what differentiated the scope of Institutional Economics from market economics. No wonder that some of the Institutionalists considered market economics superficial; they were concerned with more fundamental things. Whereas market theory was concerned with higgling over price in the market, Institutional Economics was concerned with maneuvering for position of economic advantage either inside or outside the market, ultimately through manipulating the institutional organization of society, whether by individual firms or by major social classes. These maneuvers were seen not as violations of laissez-faire or non-intervention but as in accordance with the fact of existence of both market institutions and non-market institutions of economic relevance as integral parts of the economic decision making process. Property rights, or property rights vis-à-vis other rights, for example, were part of the institutional fabric and were always subject to change. As Commons saw, in the modern Western economy, conflict over power and over opportunity for position at the banquet table often turned on property-right issues. While market theory took these rights for granted (and was often used to lend support to status quo rights' patterns, whatever they were), these rights were seen as part of that larger set of problems focusing on the problem of order, problems with processes which were active and on-going and with resolutions which were contingent rather than given or final.

So, therefore, the continuing problem of the institutional organization of the economy and the interests which that organization will serve, is the distinctive central problem of Institutional Economics as a body of knowledge. But that problem – the problem of order in economic affairs, as Spengler put it – which clearly involves the distribution of power, also has at least two other dimensions in the view of the Institutionalists: correlative to power have been psychology and knowledge. The problem of order was manifest not only in the exercise of power but also in the relation of knowledge to socio-economic policy and in the psychological or social-psychological[109] foundations of economic behavior, including power play.

With respect to psychology, it is well known that the Institutionalists were often if not typically highly critical of the hedonism or rationality assumption of market theory. The main thrusts of their arguments, from Thorstein Veblen through A. B. Wolfe to Clarence Ayres, have been: the necessity of a more sophisticated psychology; stress upon the irrational or non-rational aspects of human behavior, including the strength of custom and habit and the culturally given content of self-interest; that psychological rationality is a matter of psychic needs and not simply profit or satisfaction maximization; and that prestige and power were important components of the foundations of behavior – all as part of a general view that profit maximizing was too limited an approach to moti-vation. In contrast, market theorists generally have had the view that maximizing of profit or consumer satisfaction was an appropriate, convenient, and not inordinate methodological assumption, and that not much more psychological analysis was really necessary with which to analyze the allocation of resources and the operation of the market mechanism. Economists were less strictly concerned with wants – whether a function of pure taste, habit, or custom – than with the operation of market forces into which demand functions enter as so much data.

This controversy remains unresolved, though it has not restrained market theoreticians from the further development of analyses and models with useful results; indeed, the field is much richer what with the development of behavioral and related approaches to the theory of the firm, a partial result of Institutionalist criticism. The controversy has been somewhat more formally renewed with the publication of Galbraith's *New Industrial State* and its challenge to the motivation analysis of market theory with respect to both the businessman (or the technostructure) and the consumer. Much of this reads like replays of older controversies in which the Institutionalists were leading protagonists.

But what distinguished the Institutionalist discussion on psychology was much less their – more well known – criticisms concerning businessman and consumer motivation, though they were abundantly forthcoming, and much more their consideration of the psychological facet of the problem of order. A brief summary of this consideration will now be given, with the reminder that the purpose is not to inquire into the substance of the Institutionalists' theories per se but rather to delineate certain psychological subject-matter as problems of Institutional Economics.

First of all, it should be made clear that the Institutionalists seem, in general, to have stressed the importance of volitional and purposive activity – a volitional or deliberative choice – in one major respect and minimized it in another; in sum, not unlike many other interpreters of the problem of reason and social policy. On the one hand, as already noted in passing, the Institutionalists shared Freud's and Pareto's emphasis upon the non-rational, irrational, and/or subconscious character of human choice. This they levelled against allegedly too or overly mechanistic assumptions of market theory and the heroic implications drawn from them. On the other hand, as was seen somewhat earlier in this section, the Institutionalists generally contemplated the ascendance of deliberative over non-cognitive or non-reflective bases of social policy, at the margin emphasizing, on the basis of observation, the role of law in the revision and redirection of custom. Thus, Commons' "volitional psychology" nicely characterizes a major stand of Institutionalist thought, namely, the purposive, goal directed nature of economic policy, or activist view of social control.

This differentiation – which should not be drawn too far or too strictly – serves to put into perspective the Institutionalist critique of the psychological assumptions of market economics. For what the Institutionalists were primarily interested in was *not* the derivation of demand curves under utility maximizing assumptions but something which was to them more important, to wit: the psychological dimension of the institutional organization of society. Now, the Institutionalists generally appear to have seen the subject of psychology as one of the combination of "genuine volitional conduct" with "institutionalized personalities,"[110] with the latter in turn the combination of individual personality or psychic structure with social pressures both cultural and inter-personal. Nowhere, it appears, is there anything like a thorough or definitive statement of these questions in the Institutionalist literature, and the following is bound to contain some over-generalization. But, and this is the important point, what the Institutionalists were interested in were the processes of socialization and individuation of the individual – the creation of individual identity concomitant with the acculturation of the individual – as they related to the structure of power and the resolution of the problems of freedom and control, and continuity and change.

Thus, most discussions were in terms of the institutional dimension of personality:

> Human beings are social phenomena. Social patterns are not the logical consequents of individual acts; individuals, and all their actions, are the logical consequents of social patterns.[111]

Once again Ayres seems to use hyperbole, since social patterns, which may be analyzed independently of individuals, are themselves the product of human behavior. But as Martindale summarizes Commons:

> Individuals are not self-sufficient, independent entities; they are what they are through their participation in the institutions or going concerns of which they are members.[112]

Thus, according to one writer who is close to if not in the Institutionalist group,

> ... individuals in all societies must apply conventions and norms which are acquired from the culture by learning and which when internalized in personalities become a part of working behavior.[113]

such that the "involvement of the economic with the social order implies that" since the economic system is "a complex of cultural and institutional patterns," then,

> The individual persons who make up the society give life to these patterns by conforming with them in the relevant situations of life. They do this by filling the roles, occupying the status positions, respecting the norms, and holding the beliefs institutionalized in the social order.[114]

Looked at differently,

> What gives the typical institutions their solidarity is the emotional conditioning the community has undergone by virtue of which people get emotional satisfaction from the continuance of the accustomed situation.[115]

But what are the norms, what are the roles, and what are the patterns which successive generations will learn through acculturation? If – as Veblen taught – each individual is inculcated into the ways of the society into which he was born, and leaves it somewhat more or less different by virtue of his having been in it, such that the inculcation of a future generation is into a somewhat more or less different society, then what is that change – that difference – to be? As Ayres puts it, tersely but to the point:

> Power relationships and emotion relationships develop together.[116]

Institutional reality – the power structure – is thus a function of power play and psychological interaction; and the latter are channeled and influenced by the former. A major strand of Institutionalist writings in this respect, therefore, treats the problem of continuity and change (as well as the usages of power) in terms of differential psychological identification with this or that component of the status quo,[117] a general view parallel to that of Pareto and to some extent that of Taussig. Conformity vs. challenge[118] – itself related to attitudes on authority, which are also related to postures concerning the role of government in the economy, particularly the role of law as an instrument of social change – was among the sources of modification of social institutions; just as in a narrower but nevertheless important field, opportunity vs. scarcity consciousness was important to the germination of unionism, according to Perlman. Also, continuity vs. change depended upon power play which in turn in part reflected the direction of aggressions.[119] Psychological rationality thus governed views on the role of government and policy generally; as John M. Clark wrote, policy issues "arise out of the way in which our economic life is organized, and the needs and inequities each group feels."[120]

Finally, the Institutionalists considered the relation of knowledge to the economic process. No less than the classicists, the Institutionalists envisioned policy being ideally made on the basis of knowledge; and, indeed, the problem-solving approach of Commons and Witte is predicated upon the bringing to bear of all the knowledge requisite to the working out of a solution to a problem of policy, regardless of academic discipline. In addition, the Institutionalists recognized that one's participation in the economic process was, inter alia, upon the basis of what one considers knowledge, and, furthermore, that "Beliefs actually held are scarcely less important than what people ought to think."[121] With respect to knowledge in general, the Institutionalists also seem to have been saying, in part, that the working rules served to define values and reality as a basis for personal and collective action, and that changing ideas influenced behavior and vice versa, all as part of the continuing resolution of the problem of order.

The author cannot dwell on this dimension of the Institutionalists' subject-matter and it will have to receive relatively short attention. But the Institutionalist inclusion of the operation of knowledge as part of their general scope of relevance is perhaps most dramatically illustrated by their insistence – particularly by the Veblenians – upon technology as a prime mover in socio-economic change and organization. Technology, as they generally put it, is learned behavior with respect to skills and tools but behavior fundamentally embodying knowledge; as Galbraith more recently put it, "Technology means the systematic application of scientific or other organized knowledge to practical tasks."[122] If any single variable characterizes, e.g. Ayres' theory of economic development and socio-economic change, it is technology; no greater place could be given in his analysis to what is essentially knowledge with a power coefficient.

From all that has been written above it may be evident that the substance of no miniscule amount of Institutionalist analysis would necessarily be concerned with the development and functioning of the working rules.[123] These rules and principles – revealed in "transactions and attitudes"[124] and ensconced in custom, morality, and law –, comprise much of ordinary knowledge, and govern behavior, inter-personal relations and interaction, and conflict resolution, as well as personal judgment both emotional and deliberative. They are necessary (but not necessarily sufficient) to produce order[125] and are the heart of collective action:[126] they function to allocate scarce resources and power positions, govern the resolution of the problems of freedom and control, and continuity and change, sanction the use of power and coercion, and enter into the evolution of capitalism[127] as both cause and effect. The working rules thus govern the structure of power but much more also: for the resolution of order in all its ramifications generally takes the form of working rules, and the dynamics or evolution of the problem of order is expressed in the transformation of the working rules and thereby the power structure they both govern and reflect. Issuing from experience and nurtured by reason and interest as well as inertia, the working rules govern "the ways in which private purposes are made consistent with public purposes . . . within going concerns which restrain, liberate, and expand individual action through working rules, enforced by various sanctions."[128] Optimality for both analytical and policy purposes is a partial function of the "legal rules and practices under those rules that make it true,"[129] that is to say, the combination of individual preferences (themselves culturally conditioned) and moral and legal working rules (themselves the product of individual experience and power play) determine the substance of Pareto optimum.[130]

But perhaps most conspicuous and most important – because most relevant – of all the Institutionalist endeavors has been their persistent attention to the relation of the state to economic life and the profound and complicated inter-relations between market and legal processes. The relevant literature is enormous and includes such classics as chapter eight of Veblen's *Theory of Business Enterprise*,[131] entitled "Business Principles in Law and Economics;" Commons' *Legal Foundations of Capitalism*; Robert Lee Hale's *Freedom Through Law*;[132] J. M. Clark's *Social Control of Business* and, *inter alia*, his essay, "The Interpenetration of Politics and Economics;"[133] Walton H. Hamilton's *Politics of Industry*;[134] Richard T. Ely's *Property and Contract in their Relation to the Distribution of Wealth*;[135] and such shorter pieces as Edward A. Carlin's "Intangible Property as a Tool for Analyzing the Relationships between

Government and Private Enterprise,"[136] and Edwin E. Witte's attempt at summarization in his AEA Presidential address.[137] The Institutionalist literature considered as a whole comprises and represents one of the major social science efforts to both articulate the interrelations of legal and economic processes, and formulate a general model in terms of which those interrelations have meaning. In a very practical sense, the Institutionalists tended to define the interrelations of legal and economic processes as their most immediate general subject-matter, that is to say, to define the problem of order as a subject for research and as a body of knowledge in terms of those interrelations particularly as both legal and economic forces and institutions challenged and affected custom. No small amount of Institutionalist inquiry has centered upon law and economy as two systems of power and as two value clarification and selection processes, and their interaction, and thereby upon the concept, institution, and particular rights of property as the focal institution of joint legal and economic relevance, and as the focal point of legal and economic forces. In general also, the Institutionalists studied the various ways in which modern economic activity is grounded in law and the legal process as well as the response of the state to economic forces and interests as the latter attempted to mold the state to private economic goals. Economy is a partial function of law, and law a partial function of economy. More important, the two are practically inseparable, as politics and economics were intertwining fields of action; in particular, the state is not something outside of the economy. The saga is manifold: of private usages transformed into general laws; of "the rise of new social classes, and of their struggle for recognition;"[138] of the "collective bargaining state," or "the combined legal, economic, and political process by which individuals and various social groups adjusted their interest and objectives;"[139] and of "government . . . not something apart from the conflict; it is the sovereign power conferring economic power, and is controlled by the conflicting interests themselves."[140] The great arena of collective action was in the interplay of law and economy.

Two further facets of Institutional Economics as a body of knowledge remain to be discussed. The first necessarily has been touched upon at various points in the foregoing discussion: Institutionalism deals not only with the problem of freedom and control but also with the problem of continuity and change. As Gruchy has declared so well, Institutionalism is both holistic and evolutionary: it is interested "in studying the economic system as an evolving, unified whole or synthesis,"[141] as it "is a study of the disposal of scarce means within the framework of our developing economic system."[142] As Kuznets perceptively remarked, "The emphasis on institutions means . . . in the first place an emphasis on change."[143] The importance of this concern with change is indicated by the tendency – particularly among the Veblenians – for Institutionalism to be called "evolutionary Economics,'" harking back, of course, to Veblen's famous essay on "Why is Economics Not an Evolutionary Science?"[144] But evolutionism characterizes the group as a whole and is manifest in various and subtle ways: as in Witte's statement that,

> Institutions cannot be taken for granted, as they are man-made and changeable. Changes in the working rules are possible and occur frequently, although normally only slowly.[145]

in J. M. Clark's insight that the Institutionalist

> ... does not view the stereotype as natural and specific changes as unnatural or artificial; all forms of behavior are equally natural.[146]

and in Gambs and Wertimer's proposition that,

> ... economic systems are only bundles of institutions, and like institutions, subject to change.[147]

Analysis of the fact of change, of the genesis of changes and of the direction and control of change – particularly change in the specifics of the institutional organization of the economy but also the evolution of the system as a whole as it changes incrementally – is accordingly one of the central subjects of Institutional Economics.

Evolution in the minds of most Institutionalists connoted the gradual alteration of the structure of economy and society, including the emergence of capitalism from feudal, mercantile, and monarchical orders. They have also generally interpreted the institutional economic history of the past century and a half (and beyond) as involving the pluralization or democratization of capitalism.[148] The Institutionalists tend to picture the economy and particularly capitalism as an evolving phenomenon but specifically evolving in the direction of a broad-based polity and a broad-based economy. Not all the Institutionalists were as optimistic as Commons, nor did many others participate as actively as Commons in promoting the interests of workers through unions and the interests of the masses generally through making government responsive to their demands for action whose realization has been labelled either the service state, the regulatory state, or the welfare state. Both in terms of what Commons saw (*is* knowledge) and what Commons wanted (*ought* knowledge),

> Commons is thus the father of a labor struggle theory which is not a class struggle theory in the Marxian sense. It is not a struggle by the rising group to liquidate the old class or to raze the social structure which the latter controlled, but essaying instead to add to the old edifice new and spacious wings to serve as the dwelling places of the customs of the rising class.[149]

As Commons himself summarized centuries of evolution of the structure of society – which the present author quotes to illustrate further the evolutionary treatment of the broad question of socio-economic organization, ultimately treating the fourth basic economic problem –,

> Social institutions are in a constant change and evolution. Forms of government, of the family, of the Church, of private property are by no means the same as they were a generation ago. All these institutions originated as coercive instruments for controlling the masses and the weaker classes in the interests of the few and the strong ... The development of institutions from primitive times to the present has consisted, not in abolishing the principle of coercion but in elevating those who were suppressed into partnership with those who owned them. The family has become a cooperative association of lovers. Government and the control of industry are open to the serf and the slave ... This movement is still in progress.[150]

Finally, Institutional Economics includes a theory of value, albeit different from that of market economics. The latter has had value theory devolve into a theory of resource allocation, with factor and commodity prices serving as links between resource allocation and income distribution. In a general way, Institutionalism has acknowledged the importance of microeconomic theory, particularly as an opportunity cost theory of pricing or value,[151] so far as factor and commodity pricing in the market is concerned, though Institutionalists have criticized market economists for over-emphasizing value theory and have criticized market value theory with respect to its neglect of non-competitive conditions, the rationality assumption, over-zealous apologetics, and its neglect of factors influencing relative bargaining power. Certainly the Institutionalists have not developed any theory to rival microeconomic theory, whose descriptive accuracy has been greatly enhanced by analyses of non-competitive conditions over the past three decades or so; although in the area of distribution theory, where again Institutionalists have not developed a powerful rival body of theory, they have continued to challenge the marginal productivity theory as a complete explanation.

The preceding paragraph may be so general as to fail to give credit to various Institutionalist writings and views on price and distribution theory; no slight is intended. The basic point to which the immediate foregoing is intended to be prelude is that Institutionalists, whether or not they have learned to live with the value (price and resource allocation) theory of market economics, have really been interested in value considerations of much broader, and in their view much more fundamental, scope.

As part of the subject-matter of Institutional Economics considered as a body of knowledge, value theory to the Institutionalists has primarily meant the values and valuation process concerned with the structure of power, with the relative rights governing the power structure, and with the working rules which influence both the relative rights and the transmutation of private into social interests. That is why, when Commons inquired into and tried "to work out an evolutionary and behavioristic, or rather volitional, theory of value," he was led to the administration of reasonableness (hence Reasonable Value) by the courts – of which the Supreme Court was the final and authoritative faculty of Political Economy – and found that what he was "really working upon was not merely a theory of Reasonable Value but the Legal Foundations of Capitalism itself."[152] For, more fundamental than commodity value is social or public, or reasonable, value, whose theory encompasses the analysis and valuation of values incorporated in relative rights, conflict resolution, social order, and the power structure, and as the latter influences the former. These are the values embodied in the working rules, and thereby concerned "with the structure and proportioning of opportunities and with public policies where the objects of valuation are the working rules, which are the very structure of social organization."[153]

> In Commons analysis, . . . whatever scope society accords individual choice and valuations is a consequence of the latitude for discretionary action which is built into the system. Thus, the problem of public value, in his view, was that of determining the reasonableness of the working rules . . .[154]

The problem of value to the Institutionalists thus includes a theory of choice with respect to the opportunity costs involved in shaping and reshaping the power structure. But it is more than that. It also includes the concept that market price is relative to institutions; that, as economic laws work out differently in different institutional settings, so market "value is relative to given social institutions."[155] There needs to be a "general theory of value,"[156] including therein the value choices embodied: (a) in institutions and in institutional alterations; (b) in market valuation; (c) in the impact of institutional value upon market valuation (i.e. the impact of institutional organization – and its values – upon exchange value and resource allocation and income distribution); and (d) in the working rules which constrain Pareto optimality. The resolution of the problem of order is seen, then, as a means to valuational ends; not just values and valuation as a means to order.[157]

Institutionalist value theory is concerned with the impact of the nuances of reasonableness in conflict situations throughout economic life as a whole, particularly with respect to institutional organization (viz, the distribution of power) but also, in the context of the Veblenian concept of enhancing the life process, with respect to the impact of real living conditions and the distribution of levels of real income made possible by technology. Ultimately reasonable value, or value in terms of the life process, refers to the values embodied in the working rules governing not only the structure of power but also the allocation of resources, the level of income, and the distribution of income, particularly as the outcome is influenced by institutional organization and as the latter is influenced by private power. Social valuation, to the Institutionalist, is infinitely more encompassing than market valuation, and more encompassing than the values finding expression in the market. As J. M. Clark put it, the Institutionalists shared "a refusal to accept the market as an adequate vehicle for expressing the importance of things to society."[158] Therein lies value theory as it is distinctively Institutionalist and therein lies also the awful but challenging intractability of Institutional Economics. If the foregoing indicates what the author intends it to indicate, it should be clear that however broader the scope of Institutional Economics vis-à-vis market economics, and however broader the scope of Institutionalist value theory vis-à-vis market value theory, Institutional Economics deals with a distinctive subject-matter of truly profound importance not only to the comprehension of the economic system as a whole, including the fuller understanding of a market economy, but to the social sciences generally.

III

The purpose of this paper has been to establish the meaning of Institutional Economics as a body of knowledge in terms of a coherent set of problems. That task has been completed, at least so far as the author has been able to accomplish what has not been done fully satisfactorily (at least in the sense of wide acceptability) before, and what many have thought impossible. The fact that this has not been done before and that so many have considered it well nigh impossible have been a constant source of caution and correctly suggest restraint in expectations. The author hopes, nonetheless, that this paper is at least a step in the right direction. As already indicated, the author is, moreover, solicitous that the total historic meaning of Institutionalism resides in the combination of protest (reformism), problem solving, and the analysis of power; and that by treating Institutional Economics as a body of knowledge the author may be denying to Institutionalism part of its historic identity. The author agrees that to insist (which he does not) that it is *only* a body of knowledge would be a gross distortion; what he does insist upon is that Institutional Economics has meaning as a coherent body of knowledge and that it is capable of being interpreted as such *pro tanto*.

Assuming that the author has been successful, the meaning of Institutional Economics as a body of knowledge as articulated above may be seen as the retrospect of Institutionalism and as its legacy to the future and perhaps its prospect as well.

If the future resembles the past, then Institutionalism shall have several tasks. One of these will lie in its role as a protest movement encouraging and formulating reformation of the status quo whatever it will be. If policy over change should be on the basis of knowledge, then certainly the normative genius of Institutionalism has been Platonic in character, on the one hand serving as a utopian critic (*a la* Mannheim[159]) of existing reality, and on the other hand marshalling knowledge in constructive problem-solving efforts, somewhat in the manner (dichotomized to be sure) that Institutionalism encompassed both Veblen's pungent satire and brash castigations, and Commons' quest for policy solutions through institutional creations.

A second task should continue to be that of a constructive critic of market theory. Institutionalism may well continue to function as something of the conscience of the profession, even seemingly alienated at times, contributing an independent appraisal and critique of the positive and normative elements in market theory. Hopefully, Institutionalism may dissolve tendencies toward dogmatic and rigid orthodoxy,[160] and against what Veblen saw as the priestly function of presiding over "a highly sterilized germ-proof system of knowledge, kept in a cool, dry place."[161] Closely related, Institutionalism may continue as the haven of those who study the unconventional, of those who, for whatever reason, are outside the mainstream of a scholarship which tends to fall "into an immaterial nepotism in the topics it considers worth sponsoring,"[162] a scholarship thereby contributing to "that monumental misallocation of intellectual resources which is one of the most striking phenomena of our times."[163]

But Institutional Economics has more substance than that of critic and academic Bohemian,[164] as invaluable as its potential emancipation of others' minds and imagination may be. Institutionalists have the further and fundamental intellectual task of pursuing comprehension of their subject-matter, of studying the problems which distinguish Institutional Economics as a body of knowledge both for its own sake and to illumine policy choices. They have the task of inquiring into the institutional character of the economic system; into the resolution of all four basic economic problems, including the valuational and power processes involved in the very organization and control of the economy; into the resolution of the problems of freedom and control, and continuity and change; and, inter alia, into the interaction between market and non-market forces and institutions, including law and the market economy.

There are at present three great needs of the profession of economics as a body of knowledge: (1) the integration of micro and macro theory; (2) the construction of a general theory of the interrelation of legal and market processes in a predominantly market economy; and (3) the construction of a general theory or model incorporating both market and non-market (e.g. socialist) economies. A fourth would be the construction of a general theory of economic development, though it may be seen as a facet of each of the first three. In any event,

Institutional Economics, by virtue of its distinctive (but not unique) subject-matter, is in a position to make significant contributions in each of these areas. Moreover, the fields of comparative systems, legal-economic research, and economic development, as well as the various studies generally under the rubric of welfare economics, all are increasingly being seen as amenable to and requiring contributions such as the Institutionalists are capable of rendering. There is every reason to believe that Institutional Economics in these areas, as well as in its own distinctive subject matter, could eventually produce insights and laws of higher generalization and power than has been achieved hitherto.

But Institutionalism, still considered as a body of knowledge, has many difficulties and, indeed, dangers. One of them, of course, is the enormity and complexity of its subject-matter. Given the importance of that subject-matter, this means – though it is banal to say it – that the inquiry must be done carefully rather than carelessly, and that it must be done. The tendency for analysis to become a theory of history and a theory of society will remain, however. A second danger lies in over-intellectualization of what "is a social process with intellectual elements contained within it rather than an intellectual process . . ."[165] What is an admixture of power, psychology, and, albeit, knowledge (but in diverse forms and with radically divergent epistemological grounds) can and must be analyzed with intelligence but the process itself is neither intellectual nor mechanistic. A third danger, one of distortion also, is that of an over-emphasis upon social control. Concentration upon institutions, power, law, morals, and the like, too readily may become a preoccupation with authority and control. Yet, while it is easy to say that an emphasis upon freedom is desirable, freedom is nevertheless a function of the *pattern* of freedom and control; freedom, like power, is relational, and it is more meaningful to write of a structure of freedom than freedom, and discussion in terms of freedom tends to ignore the status quo system of control. But the danger of myopia remains. Fourth, there is the danger, inherent throughout the social sciences, of normative considerations predominating over positive analysis. The activist and reformist element in Institutionalism has somewhat tended to have this effect in the past; it would certainly remain in the future. But the difficulty – in certain ways actually a healthy sign –, as already indicated, is not generic to Institutionalism alone. Finally, there is a dual problem relating to the ambiguity of Institutionalists in the past toward technology and other institutions. On the one hand, Institutionalists have to research more carefully the relation between technology and other institutions (the lag of culture behind technology is a significant proposition but there are many other facets of, and forms to, the problem); and on the other, they have to examine much more carefully also the relation between technological possibilities and the criteria by which those possibilities are (insofar as they are or may be deliberatively) turned into actualities, that is to say, the criteria by which certain possibilities are accepted and others rejected, or more generally the criteria of weighing and selecting between opportunity costs on intensive and extensive margins. The ambiguity stems from a relative emphasis upon technological possibilities (or even imperatives, to use Galbraith's more recent phrase) over institutional restraints – a really an argument for new or different institutional arrangements –, coupled with a general emphasis upon human choice. In other words, how much of a constraint upon human choice is technology? All these difficulties, however, may be converted into opportunities for productive research and contributions to knowledge.

Perhaps the main area for the future development of Institutional Economics will lie in the area of economic development analysis. The "existing obstacles to industrial development"[166] are broadly institutional in nature, and,

> Without a theoretical framework that would establish the connections between economic and noneconomic aspects of social structure, we cannot specify [which non-economic factors bear upon underdevelopment, either as consequences or as major determining factors, or both]; and we have no such framework at hand.[167]

As Higgins has written,

> Development economists are learning that the sharp dividing lines which some of us sought to maintain between economics on the one hand, and psychology, sociology, history, anthropology, political science, and technology on the other, become fuzzy and misleading when tackling problems of economic development.[168]

> Professor Ayres and his fellows, it is now clear, have long been doing the sort of thing all "development economists" now find it necessary to do.[169]

> "If one wants to explain the economic development of a society, to make predictions as to its future progress, and to formulate long-run policy, all the social, political, historical, anthropological, and psychological factors stressed by the institutionalists must somehow be taken into account.[170]

The work of a number of recent scholars in the field[171] seems to confirm Higgins' view and to augur the recrudescence of Institutional Economic analysis – or at least the opportunity for same.

But this is only to say that Institutional Economics has been one manifestation of the Theory of Economic Policy in the profession as a whole. One could reach much the same conclusion – though with less existing achievement perhaps – in the areas of comparative economic systems and the relations of legal and economic processes in a market economy. Institutional Economics has the opportunity to contribute in its field of knowledge, a field which is viable and vibrant. If that field, and if Institutional Economics itself, does not develop as a body of knowledge – if that second of the two historic traditions in Economics does not develop its corpus of institutional theory –, then the profession and, indeed, all mankind will be so much the worse off. One may anticipate a situation in which Institutional Economics has so prospered and progressed that it too may require its independent appraisal and critique, and possibly become subject to the same criticisms which it levelled against market economics in generations past – as overly mechanistic, dogmatic, and so on. That situation would not be entirely a happy one, but as a body of knowledge Institutionalism would have come of age.

NOTES

1. "If institutional economics be broadly defined, it is practically co-extensive with economics. If narrowly defined in connection with a Veblenian origin, it consists mainly in a few thin essays, critical, hortatory, and hopeful. If not defined at all, it is a miscellaneous body of works associated with a group of economists reputed to be institutionalists." Paul T. Homan, "An Appraisal of Institutional Economics," *American Economic Review*, vol. 22 (1932), pp. 15, 16; cf. *American Economic Review, Papers and Proceedings,* vol. 22 (1932), p. 107.

2. Addison T. Cutler, "The Ebb of Institutional Economics," *Science and Society*, vol. 2 (1938), pp. 448, 461, 462, 463, 469, 470. In a subsequent note responding to the Cutler article, Joseph Dorfman wrote: "Yet it must be admitted that the label 'institutional economics' is just as vague as `Marxian economics' to many intelligent students of economic affairs." Joseph Dorfman, "On Institutional Economics," *Science and Society,* vol. 3 (1939), p. 509.

3. Kenneth E. Boulding, "A New Look at Institutionalism," *American Economic Review, Papers and Proceeding*, vol. 47 (1957), pp. 1, 3, 11–12, and passim. A more critical view was given earlier in Boulding, *A Reconstruction of Economics* (New York: Wiley, 1950), but as subsequently modified; cf. Robert A. Solo (Ed.), *Economics and the Public Interest* (New Brunswick: Rutgers University Press, 1955), p. 15.

4. Lafayette G. Harter, Jr., *John R. Commons: His Assault on Laissez Faire* (Corvallis: Oregon State University Press, 1962), pp. 249, 255.

5. The author will use the term "market economics" to refer to the main body of economic thought generally commencing with Smith and developing through such figures as Ricardo, Walras, Marshall, Keynes, Friedman, and Samuelson, consisting primarily of microeconomics and increasingly of macroeconomics; and referred to in the literature as traditional, orthodox, basic, standard, bourgeois, theoretical, neoclassical, and classical economics and sometimes as economics per se, in either or both a descriptive or pejorative sense. See note 52.

6. Edwin E. Witte, "Institutional Economics as Seen by an Institutional Economist," *Southern Economic Journal*, vol. 21 (1954); and "Economics and Public Policy," *American Economic Review*, vol. 47 (1957). Cf. Warren J. Samuels, "Edwin E. Witte's Concept of the Role of Government in the Economy," *Land Economics*, vol. 43 (1967).

7. Ben B. Seligman, *Main Currents in Modern Economics* (New York: Free Press, 1962), p. 129.

8. John S. Gambs, *Beyond Supply and Demand* (New York: Columbia University Press, 1946), p. 86.

9. Douglas F. Dowd (Ed.), *Thorstein Veblen: A Critical Reappraisal* (Ithaca: Cornell University Press, 1958), p. vii.

10. Cf. Jack E. Robertson, "Folklore of Institutional Economics," *Southwestern Social Science Quarterly*, vol. 41 (1960), p. 31 and passim; and Boulding, "A New Look at Institutionalism," *op. cit.*

11. William Jaffé, in "Economic Theory – Institutionalism: What It Is and What It Hopes to Become," *American Economic Review, Papers and Proceedings,* vol. 21 (1931), p. 140. Cf. the exchange between Robert M. Solow and John Kenneth Galbraith, "The New Industrial State: A Discussion," *The Public Interest*, no. 9 (Fall 1967).

12. Subsequent discussion in the text will make clear that reformism is not necessarily equivalent to Institutionalism any more than Institutionalism is necessarily reformism, that one can be a Platonist without being an Institutionalist and vice versa.

13. As Mitchell wrote, moreover, "When men have learned what consequences must be expected from certain operations, they can choose those leading to consequences they prefer." Quoted by Simon Kuznets in Joseph Dorfman, et al., *Institutional Economics* (Berkeley: University of California Press, 1963), p. 106. Also cf. Wesley C. Mitchell, *The Backward Art of Spending Money* (New York: Kelley, 1950), pp. 372–374 and *passim*; and Samuels, *op. cit.*, p. 146 and references cited.

14. Cf. Gambs, *op. cit.*, pp. 71–72.

15. Two further matters which the paper does *not* examine, in addition to not inquiring into the particular theories of the Institutionalist writers, are: (a) the origins of, or intellectual and social movements parallel and/or companion to, Institutionalism in Economics; and (b) the impact of and/or the reception given various Institutionalist concepts, including the ways in which Institutional ideas have been absorbed into the mainstream of Economics or the Institutionalist aspects of contemporary economic analysis. On the former, inter alia, see Boulding, "A New Look at Institutionalism," *op. cit.*; Joseph Dorfman, *The Economic Mind in American Civilization* (New York: Viking, 1959), Volume Four, and Dorfman, et al., *op. cit.*, chapter one; and Allan G. Gruchy, *Modern Economic Thought* (New York: Prentice- Hall, 1947). On the latter, cf. R. A. Gordon, in Dorfman, et al.

16. Cf. Gordon, in Dorfman, et al., *op. cit.*, pp. 139–141. See also the Solow-Galbraith exchange, *op. cit.*

17. Frank H. Knight, "Institutionalism and Empiricism in Economics," *American Economic Review, Papers and Proceedings*, vol. 42 (1952), p. 51.

18. *Ibid*, p. 54.

19. *Ibid*, p. 55.

20. John R. Commons, *The Legal Foundations of Capitalism* (Madison: University of Wisconsin Press, 1959), p. 2. Cf. Ben B. Seligman, "On the Question of Operationalism," *American Economic Review*, vol. 57 (1967), p. 155; and Dorfman et al., *op. cit.*, pp. 89, 93.

21. Inter alia, cf. Gambs, *op. cit.*, chs. 3 and 4; Fritz Karl Mann, "Institutionalism and American Economic Theory: A Case of Interpenetration," *Kyklos*, vol. 13 (1960), pp. 316ff; Dowd, *op. cit.*, ch. 17; Boulding, "A New Look at Institutionalism," *op. cit.*, pp. 8–9; Benjamin Higgins, "The Economic Man and Economic Science," *Canadian Journal of Economics and Political Science*, vol. 13 (1947), pp. 587–589, and "Some Introductory Remarks on Institutionalism and Economic Development," *Southwestern*

Social Science Quarterly, vol. 41 (1960), p. 20. *Per contra*, see Clarence E. Ayres, "Institutionalism and Economic Development," *Southwestern Social Science Quarterly*, vol. 41 (1960), pp. 57–58.

22. E. M. Burns, "Does Institutionalism Complement or Compete with 'Orthodox Economics'?," *American Economic Review*, vol. 21 (1931), p. 81. In addition, as Gordon has pointed out, "Empirical research meant different things to" Veblen, Mitchell, and Commons. Dorfman, et al., *op. cit.*, p. 126. One of the main Institutionalist points, also recognized by others, of course, is that methodological procedure is often crucially dependent *upon* the subject matter under study, that is to say, that "the degree of exactitude possible . . . is dependent on the nature of the phenomena, not on the family connections claimed for any special science," (Gambs, *op. cit.*, p. 30) such that Richard Ruggles could write that, "The advantage of institutionalism lies in its ability to deal with those factors which are not readily amenable to the present state of quantitative measurement, as well as those which can be handled statistically." Bernard F. Haley (Ed.), *A Survey of Contemporary Economics* (Homewood: Irwin, Volume Two, 1952), p. 427.

23. Cf. Ernest Nagel, "Assumptions in Economic Theory," *American Economic Review, Papers and Proceedings*, vol. 53 (1963), pp. 214ff. Compare Witte, "Economics and Public Policy," *op. cit.*, p. 12, and Seligman, "On the Question of Operationalism," *op. cit.*, p. 147, with Knight, *op. cit.*, p. 45.

24. John S. Gambs and Sidney Wertimer, Jr., *Economics and Man* (Homewood: Irwin, 1959), p. 138; Seligman, "On the Question of Operationalism," *op. cit.*, and A. W. Coats, "The Influence of Veblen's Methodology," *Journal of Political Economy*, vol. 62 (1954).

25. Dorfman et al., *op. cit.*, pp. 90, 98ff, 108ff, 133ff and *passim*.

26. Burns, *op. cit.*, p. 85.

27. Joseph J. Spengler and William R. Allen (Eds), *Essays in Economic Thought* (Chicago: Rand McNally, 1960), p. 492.

28. Boulding, "A New Look at Institutionalism," *op. cit.*, pp. 7, 8.

29. Dorfman, et al., *op. cit.*, p. 90.

30. Ben B. Seligman, Book Review, *American Sociological Review*, vol. 28 (1963), p. 490; cf. Seligman, *Main Currents in Modern Economics*, *op. cit.*, p. 178.

31. Cf. Gambs, *op. cit.*, p. 53; and H. L. McCracken, "A Supplement to Dr. Witte's Paper on Institutional Economics," *Southern Economic Journal*, vol. 22 (1955), p. 104.

32. F. S. C. Northrop, *The Logic of the Sciences and the Humanities*, (Cleveland: World, 1959), p. 270.

33. Robert L. Heilbroner, *The Worldly Philosophers* (New York: Simon and Schuster, Third Edition, 1967), p. 208.

34. Cf. David Hamilton, "Why is Institutional Economics Not Institutional?," *American Journal of Economics and Sociology*, vol. 21 (1962), and "Veblen and Commons: A Case of Theoretical Convergence," *Southwestern Social Science Quarterly*, vol. 34 (1953); Mann, *op. cit.*; Walton H. Hamilton, "The Institutional Approach to Economic Theory," *American Economic Review*, vol. 9 (1919); and Gambs and Wertimer, *op. cit.*, p. 191.

35. Witte, "Institutional Economics as Seen by an Institutional Economist," *op. cit.*, p. 135 and passim.

36. Witte, *ibid*, pp. 135, 137.

37. Cf. Gambs, *op. cit.*, pp. 82–83; Burns, *op. cit.*, pp. 80, 83, 85; Lionel D. Edie, "Some Positive Contributions of the Institutional Concept," *Quarterly Journal of Economics*, vol. 41 (1927), pp. 411, 420, 434; McCracken, *op. cit.*, p. 106; Richard W.

Taylor (Ed.), *Life, Language, Law* (Yellow Springs, Ohio: Antioch Press, 1957), p. 87; Paul T. Homan, "Economics: The Institutional School," *Encyclopaedia of the Social Sciences* (New York: Macmillan, 1935), Volume 5, p. 390; and Stow Persons (Ed.), *Evolutionary Thought in America* (New York: George Braziller, 1956), wherein Spengler writes that, "Yet, probably because institutionalists lacked a coordinating body of theory, it did not give rise to a school." (p. 243) See note 38.

38. Harter wrote that, "The decline of institutionalism has been the result of more than the appropriation of its methodology and subject matter by the main stream of economics and sociology. The greatest cause of the decline of the school has stemmed from the centrifugal nature of its practical bias." (*Op. cit.*, pp. 252–253; cf. p. 254).

39. Ayres, *op. cit.*, pp. 59–62; and Gruchy, *op. cit.*, passim.

40. Witte, "Institutional Economics as Seen by an Institutional Economist," *op. cit.*, p. 137; and Taylor, *op. cit.*, pp. 91–92.

41. Even a general body of Institutional theory would be "independent of particular institutions or cultures," at least to some extent. Higgins, "The Economic Man and Economic Science," *op. cit.*, p. 594.

42. Cf. Knight, *op. cit.*, pp. 47, 49.

43. Cp. Manuel Gottlieb, "The Theory of an Economic System," *American Economic Review, Papers and Proceedings*, vol. 43 (1953), pp. 353–354; and Spengler and Allen, *op. cit.*, p. 498.

44. Cf. Mann, *op. cit.*, p. 319; Harter, *op. cit.*, p. 250; and David Hamilton, "Why is Institutional Economics Not Institutional?," *op. cit.*, p. 313.

45. Inter alia, cf. C. E. Ayres, "Ideological Responsibility," *Journal of Economic Issues*, vol. 1 (1967), and Louis J. Junker, "Capital Accumulation, Savings-Centered Theory and Economic Development," *ibid, loc. cit.*

46. Cf. Mitchell, *op. cit.*, pp. 336ff, 362; Dorfman, et al., *op. cit.*, pp. 89–90; Seligman, Book Review, *op. cit.*, p. 153; and Boulding, "A New Look at Institutionalism," *op. cit.*, p. 12.

47. See Warren J. Samuels, "The Nature and Scope of Economic Policy," in Samuels, *The Classical Theory of Economic Policy* (Cleveland: World, 1966), appendix.

48. Gruchy has stressed that the question of scope "should be determined, not by the desire for measurement or exactitude, but by the desire for a full understanding or interpretation of the flow of economic events. The limits of the scope of economic science should not be set by any methodological preference, as in the case of those who refuse to go beyond precise or quantifiable data, but rather by the nature of the problems encountered in actual economic experience." (Allen G. Gruchy, "Issues in Methodology: Discussion," *American Economic Review, Papers and Proceedings*, vol. 42 (1952), p. 68.) As Spengler has pointed out, "Mill believed that economists exaggerated the effect of competition at the expense of custom; but he added that 'only through the principle of competition has political economy any pretension to the character of a science.'" (Spengler and Allen, *op. cit.*, p. 32.) Using Spengler's terminology, to limit economics to competition so as to make it "scientific" is to limit it to one definition of a hypothetical subrealm. In part, the problem of scope is a question of taste; but it also involves the controversy as to whether scope is to be a function of research techniques or of the problem studied per se. Both the choice of research technique and the choice of problem are normative. What this reduces to is that economic analyses and models are comprised of varying combinations of intensive and extensive margins.

49. Edie quoted J. B. Clark's *Philosophy of Wealth* that, "The ultimate foundations of political economy lie deeper than the strata on which existing systems have been reared." Edie, *op. cit.*, pp. 438–439.

50. "Nor is institutionalism in conflict with every aspect of classical economics. Marshall, Mill, and Ricardo told part of the institutional story in occasional paragraphs, sentences, or footnotes. To the modern institutionalist, the classical footnote becomes a book, and the classical book becomes a footnote needing revision." Willard E. Atkins, "Institutional Economics: Round Table Conference," *American Economic Review, Papers and Proceedings,* vol. 22 (1932), p. 112.

51. See Samuels, *The Classical Theory of Economic Policy, op. cit.*, Chapter One and Appendix.

52. Herein derives part of the terminological problem with respect to Institutional vis-à-vis market economics. Institutional Economics can be said to be equivalent to Theory of Economic Policy; *or* Institutional Economics can be said to be equal to the sum or combination of market economics and Theory of Economic Policy, in which case Institutional Economics *is* Economics. The present author here uses Economics to be the more inclusive category, with Institutional Economics practically equivalent to Theory of Economic Policy, such that Economics is equal to the sum or combination of market and Institutional Economics (with Institutional Economics equal to Theory of Economic Policy). In the following sections, however, the term Institutional Economics will refer only to its traditional meaning, that is to say, to include only the Institutionalists (always connoting Veblen, Commons, Mitchell, Witte, Ayres, et al.) and not the non-Institutionalist theorists of economic policy (Knight, Hayek, Adam Smith, et al.): while the author feels that the basic problems dealt with by Institutional Economics (the Institutionalists) are essentially the same as those dealt with by the theorists of economic policy of a non-Institutionalist genre, the scope of the present paper is limited to the Institutionalists.

53. Inter alia, see Samuels, *ibid*; Lionel Robbins, *The Theory of Economic Policy in English Classical Political Economy* (London: Macmillan, 1953); Spengler and Allen, *op. cit.*, pp. 11–25; and John R. Commons, *Institutional Economics* (Madison: University of Wisconsin Press, Two Volumes, 1959), particularly Volume One.

54. Inter alia, see W. J. Baumol, *Welfare Economics and the Theory of the State* (Cambridge: Harvard University Press, Second Edition, 1965), pp. 1–48; Morris A. Copeland, "Institutionalism and Welfare Economics," *American Economic Review*, vol. 48 (1958), p. 2 and passim; Gerald G. Somers (Ed.), *Labor, Management, and Social Policy* (Madison: University of Wisconsin Press, 1963), p. 6; and Burns, *op. cit.*, p. 81.

55. Dorfman refers to Commons "as a pioneer in the theory of economic policy." Joseph Dorfman, "The Foundations of Commons' Economics," in John R. Commons, *The Distribution of Wealth* (New York: Kelley, 1963), p. xv. Cf. Taylor, *op. cit.*, p. 85.

56. See Spengler and Allen, *op. cit.*, pp. 1ff; Gambs, *op. cit.*, pp. 8, 14; Witte, "Institutional Economics as Seen by an Institutional Economist," *op. cit.*, pp. 133, 135; Knight, *op. cit.*, p. 53, cf. pp. 49–50; Taylor, *op. cit.*, pp. 86, 90; and Gottlieb, *op. cit.*, pp. 351–352, 354, 355.

57. Mitchell, *op. cit.*, p. 373.

58. *Ibid*, p. 336.

59. Cutler, *op. cit.*, p. 449; Solo, *op. cit.*, p. 18; Seligman, *Main Currents in Modern Economics, op. cit.*, p. 146; and Gruchy, *Modern Economic Thought, op. cit.*, p. 68.

60. Mitchell, quoted in Cutler, *op. cit.*, p. 457; and Burns, *op. cit.*, p. 86.

61. Gambs and Wertimer, *op. cit.*, p. 170.

62. John R. Commons, "Materialist, Psychological, Institutional Economics," in *Economic Essays in Honour of Gustav Cassel* (London: Allen & Unwin, 1933), p. 101; and *The Economics of Collective Action* (New York: Macmillan, 1950), p. 21.

63. Joseph Dorfman, "On Institutional Economics," *op. cit.*, p. 510.

64. Solo, *op. cit.*, p. 20.

65. Thorstein Veblen, *Essays in our Changing Order* (New York: Viking, 1954), p. 43.

66. John R. Commons, *A Sociological View of Sovereignty* (New York: Kelley, 1965), p. ix.

67. Veblen, quoted in Mann, *op. cit.*, p. 309.

68. Cf. Morris A. Copeland, *Our Free Enterprise Economy* (New York: Macmillan, 1965), pp. 136–139 and passim; and, inter alia, Gottlieb, *op. cit.*, pp. 357ff.

69. George W. Stocking, "Institutional Factors in Economic Thinking," *American Economic Review*, vol. 49 (1959), p. 19.

70. Gambs, *op. cit.*, p. 90; and cf. Dorfman, *The Economic Mind in American Civilization, op. cit.*, vol. 4, pp. 375–376.

71. For example, see Abraham Hirsch, "Wesley Clair Mitchell, J. Laurence Laughlin, and the Quantity Theory of Money," *Journal of Political Economy*, vol. 75 (1967).

72. Mitchell, as quoted from Dorfman, *The Economic Mind in American Civilization, op. cit.*, vol. 4, p. 362.

73. Mitchell, *op. cit.*, p. 371. Cf. Dorfman *et al.*, *op. cit.*, pp. 96–97.

74. Mitchell, *op. cit.*, p. 371.

75. Cf. David Seckler, "The Naivete of John R. Commons," *Western Economic Journal*, vol. 4 (1966).

76. Commons, *Legal Foundations of Capitalism, op. cit.*, p. 2.

77. On the importance of non-deliberative custom as recognized by the institutionalists, see, inter alia, Ayres, *The Industrial Economy, op. cit.*, p. 47; Jaffé, *op. cit.*, p. 139; David Hamilton, "Veblen and Commons: A Case of Theoretical Convergence," *op. cit.*, p. 45; Spengler, in Persons, *op. cit.*, p. 253; Mitchell, *op. cit.*, 321, 326 and passim; Don Martindale, *The Nature and Types of Sociological Theory* (Boston: Houghton Mifflin, 1960), p. 403; Somers, *op. cit.*, pp. 13, 15, 20; Seligman, *Main Currents in Modern Economic, op. cit.*, pp. 137, 139; Richard T. Ely, "Institutional Economics: Round Table Conference," *American Economic Review, Papers and Proceedings*, vol. 22 (1932), p. 115; Atkins, *ibid*, p. 111; Harter, *op. cit.*, p. 245; and Commons, *A Sociological View of Sovereignty, op. cit.*, p. x, and *The Economics of Collective Action, op. cit.*, chapter nine and *passim*. On the recognition of the incremental and deliberative importance of the legal process and of designed reform, see Dorfman, *The Economic Mind in American Civilization, op. cit.*, vol. 4, p. 361; Gruchy, *Modern Economic Thought, op. cit.*, pp. 561–562; Dowd, *op. cit.*, p. 12; Mitchell, *op. cit.*, pp. 337, 342 and passim; Gambs, *op. cit.*, pp. 23, 39, 87; Spengler and Allen, *op. cit.*, pp. 7, 11; and Witte, "Institutional Economics as Seen by an Institutional Economist," *op. cit.*, pp. 134–135. In re-continuity with the classicists, cf. Samuels, *The Classical Theory of Economic Poling, op. cit.*, pp. 221–223.

78. Cf. Commons, *The Legal Foundations of Capitalism, op. cit.*, chapters six, seven, and eight.

79. Clarence E. Ayres, "Institutional Economics: Discussion," *American Economic Review, Papers and Proceedings*, vol. 47 (1957), p. 26; cf. Ayres, "Institutionalism and Economic Development," *op. cit.*, pp. 52, 56.

80. Ayres, *The Industrial Economy*, *op. cit.*, p. 349.

81. William N. Loucks, *Comparative Economic Systems* (New York: Harper & Row, Seventh Edition, 1965), p. 11.

82. Joseph A. Schumpeter, *History of Economic Analysis* (New York: Oxford University Press, 1954), p. 34.

83. Dorfman et al., *op. cit.*, p. 142.

84. Commons, *The Distribution of Wealth*, *up. cit.*, p. xii.

85. Dorfman et al., *op. cit.*, pp. 70, 72, 75.

86. Mitchell, *op. cit.*, p. 345.

87. Copeland, *Our Free Enterprise Economy*, *op. cit.*, pp. v, vi, 7, 14, 15, and passim, and "Institutionalism and Welfare Economics," *op. cit.*, pp. 2, 13.

88. Robert L. Hale, "Coercion and Distribution in a Supposedly Non-Coercive State," *Political Science Quarterly*, vol. 38 (1923); and other references given in Warren J. Samuels, "Legal-Economic Policy: A Bibliographical Survey," *Law Library Journal*, vol. 58 (1965), p. 237.

89. Dowd, *op. cit.*, p. 311.

90. Gambs, *op. cit.*, pp. 22–23; cf. pp. 12, 13, 16 and passim.

91. Gruchy, *Modern Economic Thought*, *op. cit.*, p. viii.

92. *Ibid*, p. 3.

93. *Ibid*, p. 26; cf. pp. 553–554 and passim.

94. Mitchell, *op. cit.*, p. 372.

95. Ayres, *The Industrial Economy*, *op. cit.*, p. 48.

96. Cf. Mann, *op. cit.*, p. 323; and Gordon, in Dorfman et al., *op. cit.*, p. 146 (but compare pp. 140–141).

97. Boulding, *A Reconstruction of Economics*, *op. cit.*, p. 5.

98. Cf. Harter, *op. cit.*, p. 242.

99. Talcott Parsons, *The Structure of Social Action* (New York: McGraw-Hill, 19373; *The Social System* (Glencoe: Free Press, 1951); with Neil J. Smelser, *Economy and Society* (Glencoe: Free Press, 1956); *Structure and Process in Modern Societies* (Glencoe: Free Press, 1960); with Edward A. Shils (Eds), *Toward a General Theory of Action* (New York: Harper & Row, 1962); and *Societies: Evolutionary and Comparative Perspectives* (Englewood Cliffs: Prentice-Hall, 1967).

100. Alfred Kuhn, *The Study of Society: A Unified Approach* (Homewood: Irwin, 1963). See also C. Addison Hickman and Manford H. Kuhn, *Individuals, Groups, and Economic Behavior* (New York: Dryden Press, 1956).

101. Robert A. Solo, *Economic Organizations and Social Systems* (Indianapolis: Bobbs-Merrill, 1967). Inter alia, see also Robert A. Dahl and Charles E. Lindblom, *Politics, Economics and Welfare* (New York: Harper, 1953).

102. Allan G. Gruchy, "Institutional Economics: Discussion," *American Economic Review, Papers and Proceedings*, vol. 47 (1957), p. 13. See William C. Mitchell, "The Shape of Political Theory to Come: From Political Sociology to Political Economy," *American Behavioral Scientist*, vol. 11 (November-December, 1967) for an example of a similar but not altogether parallel movement which has some of its origins in the studies of major theorists of economic policy who are economists but not Institutionalists.

103. Seligman, *Main Currents in Modern Economics*, *op. cit.*, p. 157.

104. Jaffé, *op. cit.*, p. 140.

105. Spengler and Allen, *op. cit.*, p. 9.

106. Samuels, *The Classical Theory of Economic Policy*, *op. cit.*, Chapter One and Appendix, and passim.

107. Spengler and Allen, *op. cit.*, pp. 22–25 and *passim.*

108. John M. Clark, *Social Control of Business* (New York: McGraw-Hill, 1939), p. 12. Cf. Samuels, "Edwin E. Witte's Concept of the Role of Government in the Economy," *op. cit.*, p. 134.

109. Gruchy, *Modern Economic Thought*, *op. cit.*, pp. 19, 563, 619.

110. Knight, *op. cit.*, p. 83.

111. Clarence E. Ayres, "The Co-ordinates of Institutionalism," *American Economic Review, Papers and Proceedings*, vol. 41 (1951), p. 49.

112. Martindale, *op. cit.*, p. 402.

113. Gottlieb, *op. cit.*, p. 356.

114. *Ibid*, p. 352.

115. Ayres, *The Industrial Economy*, *op. cit.*, p. 46.

116. *Ibid*, p. 302.

117. Cf. Taylor, *op. cit.*, p. 81; Dorfman et al., *op. cit.*, p. 57; and Gambs, *op. cit.*, pp. 34, 47, 70, 71.

118. Taylor, *op. cit.*, p. 81.

119. Cf. Gambs, *op. cit.*, p. 89; and Gambs and Wertimer, *op. cit.*, p. 180.

120. John Maurice Clark, *Economic Institutions and Human Welfare* (New York: Knopf, 1957), p. 234.

121. Witte, "Economics and Public Policy," *op. cit.*, p. 15.

122. John Kenneth Galbraith, *The New Industrial State* (Boston: Houghton Mifflin, 1967), p. 12.

123. The classic statement is to be found in Commons, *Legal Foundations of Capitalism*, *op. cit.*, pp. 134–142.

124. Dorfman, *The Economic Mind in American Civilization*, *op. cit.*, vol. 4, p. 388.

125. Spengler and Allen, *op. cit.*, pp. 24ff, and notes 4 and 78.

126. Wesley Mitchell, *op. cit.*, pp. 320–321.

127. Sherman Roy Krupp (Ed.), *The Structure of Economic Science* (Englewood Cliffs: Prentice-Hall, 1966), p. 187.

128. Somers, *op. cit.*, p. 6.

129. Copeland, "Institutionalism and Welfare Economics," *op. cit.*, p. 4.

130. See Samuels, *The Classical Theory of Economic Policy*, *op. cit.*, pp. 250, 272, particularly the references to James Buchanan and Oskar Morgenstern.

131. New York: Scribner's, 1904.

132. New York: Columbia University Press, 1952.

133. Clark, *Economic Institutions and Human Welfare*, *op. cit.*, chapter 10.

134. New York: Knopf, 1957.

135. New York: Macmillan, Two Volumes, 1914.

136. *Quarterly Journal of Economics*, vol. 67 (1953).

137. Witte, "Economics and Public Policy," *op. cit.*, pp. 4–11.

138. Martindale, *op. cit.*, p. 403.

139. Dorfman, *The Economic Mind in American Civilization*, *op. cit.*, vol. 4, p. 390.

140. Bushrod W. Allin, "Is Group Choice a Part of Economics?," *Quarterly Journal of Economics*, vol. 67 (1953), p. 366.

141. Gruchy, *Modern Economic Thought*, *op. cit.*, p. viii.

142. Gruchy, "Institutional Economics: Discussion," *op. cit.*, p. 13.

143. Dorfman et al., *op. cit.*, p. 98.

144. *Quarterly Journal of Economics*, vol. 12 (1898).

145. Witte, "Institutional Economics as Seen by an Institutional Economist," *op. cit.*, p. 134.

146. J. M. Clark, "Institutional Economics: Round Table Conference$, *American Economic Review, Papers and Proceedings*, vol. 22 (1932), p. 105.

147. Gambs and Wertimer, *op. cit.*, p. 172.

148. See generally, Wesley C. Mitchell, *Types of Economic Theory* (New York: Kelley, Volume One, 1967).

149. Selig Perlman, in Henry William Spiegel (Ed.), *The Development of Economic Thought* (New York: Wiley, 1952), p. 411.

150. Commons, *A Sociological View of Sovereignty*, *op. cit.*, p. x.

151. Commons, "Materialistic, Psychological, Institutional Economics," *op. cit.*, p. 92.

152. Commons, *The Legal Foundations of Capitalism*, *op. cit.*, pp. vii, viii.

153. Kenneth H. Parsons, "institutional Economics: Discussion," *American Economic Review, Papers and Proceedings*, vol. 41 (1951) pp. 82–83.

154. Kenneth H. Parsons, "Institutional Economics: Discussion," *American Economic Review, Papers and Proceedings*, vol. 47 (1957), p. 26.

155. Edie, *op. cit.*, p. 421.

156. *Ibid*, p. 420; and McCracken, *op. cit.*, p. 106.

157. Cf. Spengler and Allen, *op. cit.*, pp. 23, 25, 491.

158. Clark, *Economic Institutions and Human Welfare*, *op. cit.*, p. 58.

159. Cf. Gambs, *op. cit.*, pp. 69–72; also pp. 12–13.

160. Joseph Dorfman, "Institutional Economics: Discussion," *American Economic Review, Papers and Proceedings*, vol. 41 (1951), p. 81; and Edie, *op. cit.*, p. 415.

161. Quoted in David Riesman, *Thorstein Veblen: A Critical Interpretation* (New York: Scribner's, 1953), p. 99.

162. *Ibid*, p. 110.

163. Kenneth E. Boulding, "The Economics of Knowledge and the Knowledge of Economics," *American Economic Review, Papers and Proceedings*, vol. 56 (1966), p. 12.

164. Riesman, *op. cit.*, p. 106.

165. Raymond A. Bauer, "Social Psychology and the Study of Policy Formation," *American Psychologist*, vol. 21 (1966), p. 935.

166. Alexander Gerschenkron, *Economic Backwardness in Historical Perspective* (New York: Praeger, 1962), p. 8.

167. Simon Kuznets, *Modern Economic Growth: Rate, Structure and Spread* (New Haven: Yale University Press, 1966), p. 437.

168. Higgins, "Some Introductory Remarks on Institutionalism and Economic Development," *op. cit.*, p. 15.

169. *Ibid*, p. 16.

170. Higgins, "The Economic Man and Economic Science," *op. cit.*, p. 599.

171. Cf. Irma Adelman and Cynthia Taft Morris, *Society, Politics & Economic Development* (Baltimore: Johns Hopkins Press, 1967); Edward C. Banfield, *The Moral Basis of a Backward Society* (Glencoe: Free Press, 1958); Everett E. Hagen, *On the Theory of Social Change* (Homewood: Dorsey Press, 1962); and many others.

MISCELLANEOUS MATERIALS

Martin G. Glaeser's
AUTOBIOGRAPHICAL NOTES

Edited by Warren J. Samuels

Martin G. Glaeser (August 11, 1888–March 19, 1967) was born in Tepliwodau, Germany. He came to the United States in 1892 and became a naturalized citizen in 1897. He received his B.A. (1911) and M.A. (1916) from Wisconsin and his Ph.D. (1925) from Harvard. He worked for the Wisconsin Railroad Commission as a statistician and case investigator during 1911–1918 (the Railroad Commission in 1931 – during the administration of Governor Philip LaFollette – became the Public Service Commission, as a result of legislation that Glaeser helped draft; that and subsequent legislation increased the authority and financial support of the Commission). He taught and practiced public utility economics at the University of Wisconsin during the periods 1917–1918 and 1919–1959. He was associated with the Tennessee Valley Authority during 1933–1938, first as a special adviser and later as chief power planning engineer.

Glaeser continued and enhanced the tradition established by Richard T. Ely in what initially was called land and public utility economics, and originally reinforced by John R. Commons through his wide-ranging reform activities in administrative regulation and scholarly work on the legal foundations of capitalism. Glaeser's early Outlines of Public Utility Economics (New York: Macmillan, 1927) was followed thirty years later by Public Utilities in American Capitalism (New York: Macmillan, 1957). The latter book reveals Glaeser's comprehensive approach to his subject. It covers the history of public utilities; the several epochs of public utility regulation; the legal, institutional and ideational history of the public utility concept and practice; the administrative

Documents on Modern History of Economic Thought, Volume 21-C, pages 253–270.
© 2003 Published by Elsevier Science Ltd.
ISBN: 0-7623-0998-9

organization of public utility regulation; the historical array of regulatory policies and their related economic theories; the technology of production of utility output; and the variety of local, state and national projects in the field of water utilities, especially the Tennessee Valley Authority. Throughout one finds the influence of Commons, including the emphasis on legal history, the theory and practice of going concerns, and a positive analysis of the economic role of government. Indeed, Glaeser's work on the legal-economic public utility institution was a significant contribution to the Wisconsin tradition in the field of the economic role of government/law and economics.

Strictly speaking, the autobiographical notes published below for the first time pertain more to the field of economic history than the history of economic thought, though, being an account of his early years, they do provide some insight into the intellectual formation of one of the foremost specialists in his field in his day – and one of the foremost trainers of the next generation of specialists. These notes are the only completed portion of Glaeser's autobiography. They were provided by Glaeser's student and my colleague, Harry M. Trebing. Glaeser was Trebing's major professor; I studied institutional economics with him.

The notes provide a brief but vivid picture of the family life of a tinsmith in Silesia, the attractions of relocation to the United States, and the insecure marginal life of a family trying in various ways to make a go of it in their adopted country. We see a glimpse of class structure in the old world and the opprobrium attached to being called a "socialist." We read the spelling "Glaser" and wonder if at Allis Island it became "Glaeser" or if the difference is only a typographical error. We sense the gradual transition from more or less self-sufficient family production to the purchase of products on the market, providing new, if insecure jobs: Raw coffee beans requiring home roasting and the consequent need for roasters – an opportunity for one with the skills of a tinsmith. We appreciate the risk-diversifying strategy of combining farm and industrial work – in a rapidly industrializing economy – and the prospect of domestic primitive accumulation of capital through home ownership. We view a family relocating to where the jobs are and the gradual erosion of old-world religious ties. We learn that not all conflicts between employer and employee are matters of socioeconomic class. We better understand an economy system in which everyone "had to find their place in the new economy" and a young man looking "forward to a better type of employment" than his father's. We appreciate how the immigrants soon began to think of themselves as Americans. And we learn of the ways in which the young Martin Glaeser became part of academia.

We smile at the machinations and varying fortunes of young men as they travelled to Europe in 1911 to visit the old world and relatives. We get a sense of German monarchism and militarism.

The document is 18 pages long plus a title page, typed double-spaced. Only minor typographical corrections have been made.

I am indebted to Harry Trebing, William Dodge and others for help in preparing this document. See, for example, "Memorial Resolution of the Faculty of the University of Wisconsin on the Death of Emeritus Professor Martin G. Glaeser," Faculty Document 130; May 1, 1967.

AUTOBIOGRAPHICAL NOTES

Martin G. Glaeser

These notes are based upon my father's and mother's memories, my own memories, my own recollections and such readings as were apposite.

I was born on August 11, 1888 in Tepliwodau, Province of Silesia, Germany. The village has since been given a German name, Lauenbrunnen. It is a small farming village, not far from the capital city of Breslau. Breslau is perhaps more important as the seat of a mediaeval university, famous as the Alma Mater of the astronomer, Kepler. My first visit to Breslau was in 1911, the year of my graduation from the University of Wisconsin and the 500th anniversary of Breslau.

Most of the memory of my birthplace depends upon the memory of my parents. Father was an important person because he was the only tinsmith serving this agricultural area which was dependent upon the requirements of large scale farmers. They were known as dominia or guthsbesitzer, and had a political reputation as Junkers. My mother's status was that of a servant girl who met my father while he was supplying some Junker's establishment with a tin roof. Father kept a tin-shop with the necessary tools and a small store arranged to give an apprentice, who lived with the family, selling experience. It was this apprentice who taught me how to sing German songs.

My socialization was limited to the daughter of a gardener, living across the street, whose tenants we were. He had developed an apple orchard around the shop and one important aspect of our activities was to use a whip as a means of appropriating apples with the two appropriators concealed in the high grass of the orchard. One other item of these early contacts should be mentioned. In the shop, father also kept a piece of working equipment – a metal boring machine, which was one of his beginnings of mechanization. It was used by him to bore holes in steel facilities, but by us as a plaything. It was later transported by him to the United States, and came to be used by him in the manufacture of coffee roasters in Sheboygan during the period of unemployment and depression from 1893 to 1894.

One important incident should be recalled as associated with my early days in Germany. This was associated with an occasional trip from Lauenbrunnen to a manufacturing village named Peterswaldau. This place was about five miles by rail from a small village that bore the religious name of Gnadenfiei and to which it was our custom to walk the two miles of shaded cherry trees. Peterswaldau was the home of my mother's relatives, her mother and father and three sisters. All of them, as was the custom, were linen weavers. The father was ill with tuberculosis, while the mother was a hard-working commercial distributor who made her way over the Owl mountains, which were the foothills of the Riesen Gebirge. The train was of real assistance although the red-capped station master was a cause of alarm to me because his ringing of the bell betokened the arrival of a giant snorting monster. At the end of the trip, however, was the promise of playthings provided by a cousin in terms of French and German soldiers which could be marched back and forth.

I did not know until later that the industrialized workers were subject to industrial ills, especially those employed by a plant owned by the firm called the Zwanzigers. This industrial concern was later made famous by being made the subject of an industrial play by Gerhard Hauptman, "The Weavers."

I should also here record the fact that my mother was unwilling to leave Germany and her relatives until father was able to convince a friend (whom we all called Tante) to join us with the promise of effective workmanship.

Father always kept some connections with his villagers, especially through the well-known central exchange, the Schenke, the village tavern. Thus, he kept some political tie-up with what was known as the Fortschritts Partei (Progressive Party). He also continued some long distance communication with former friends and trade union members. One in particular, a baker, had written from Sheboygan, Wisconsin, that favorable employment was available in America, especially for tradesmen like him. It was as a result of some local criticism that the word had been passed among father's employers that "Klempnermeister Glaser ist ein sozialist." It seems that the effect of this opinion brought about some reduction in his Junker employment with the result that writing to his Sheboygan friend, father revealed that he planned to emigrate with his family from Bremen to Sheboygan.

Unfortunately, cholera had broken out in Bremen, and father was forced to unpack his equipment and start work once more. Only later did he learn that it would have been possible to emigrate by using the Oder River and to leave from Stettin and Swinemunde by way of the Baltic Sea. All I can recall about our departure is that father sat with me in the rear of a small boat sailing from Stettin to Swinemunde, feeding gulls. When we reached our main ship, called the Gotia, we boarded the Gotia with bearded seamen watching us climbing the gangplank. All the passengers lived and slept in the hold where the ship had formerly carried cattle to West Africa. Consequently the feeding and care were most primitive. Our transport route followed the Baltic Sea, the Kattegat, the Skagerrak, and out to the North Sea. We sailed north of the Orkney Islands, crossed the Atlantic Ocean and landed at Allis Island. The crossing of three weeks was partly by steam but mostly by sails. There were times when I was permitted on deck and played with some rubber balls which were from father's shop, and most of which I lost overboard. Father, one of the younger men, had been assigned to fetch all meals in big copper kettles from the galley as a joint central eating establishment. I also recall a funeral at sea.

There were seasonal storms at sea which required father and others so delegated to take particular care in supplying soup and food so as to keep their balance. Father did have difficulty in keeping soup out of his beard.

The Gotia landed in New York (Allis Island) late in November. We were not permitted to disembark because we were first required to pass through customs and a complete program of disinfection and fumigation. Now that the original island has disappeared and has become instead a romantic survival of a shipping port, it is difficult to recall the accompanying scenes. In our case we were temporarily required to sojourn on the icy deck while the processes of fumigation were completed. Our reception in America was completely frigid. The water journey having been completed, I recall our leaving by crowded immigrant train and we became a part of the great American Railroad system which hauled us north through Canada to Chicago. Then transfer to the C and NW Railway for Sheboygan must have been accomplished without further personal disturbance. What thoughts may have been in my mind as the succession of city lights and changing municipal structures bore in upon my consciousness, it is not possible to imagine. In any event something new was transpiring which would require great reorientation, I recall nothing of the long journey by way of Niagara Falls and Chicago. It was not until Sheboygan was reached that a reaction, both recalled and induced by later reflection, caused a new response. Tepliwoda and Peterswaldau were not much as towns with common carriers. But in this new world opening up to me, horses might not be new contrivances but this was my reaction. Expressed in German, according to my father's memory, my first reaction to my future home was. "Haben die aber kleine Pferde hier?" Do they have small horses here? Sheboygan had not yet entered the electric transit car. My childish mind transmuted the use of mules into small sized horses.

No assignment seemed more apposite than the removal of the Glaeser family to its new home in America. The situs, so assigned is still available – Jefferson Avenue, the only difference being that the upstairs consisting of a kitchen and several sleeping quarters were available for tenancy. The downstairs occupancy was characteristically available to a carpet-weaving establishment whose heavy thump, thump of the weaving machinery provided its own rhythm. Mr. Liscker, a German with a heavy brogue, was the proprietor. This occupancy was subject to dispute by the infiltrating snow and ice, the kind of dispute with which father was well versed to compete. Other incidents need to be recalled with Christmas in the offing. I was sent to school at once, a reasonable distance of six blocks. Except for delays encountered by examining stores en route, there were no problems. I recall that I was left to struggle with strange teacher assignments in kindergarten. Another item was that mother, with her Christmas spirit, invaded the Bodenstein clothing store to come forth with surplus Christmas tree branches which the wily tinner was somehow able to fasten onto a broom handle as a Christmas tree.

Our stay on Jefferson Avenue lasted until midwinter when we removed to a new home on Center Street where more space was available. The reason for removal was that father had been unable to find continuous employment. Although Sheboygan had passed the pioneering and construction stage, the current depression had left it with a high margin of unemployment. Mr. J. A. Koepsell, a master employer, who would have given a job if he could, suggested removal to Milwaukee, the neighboring city. Here father was able to find a job, but it was not of over-long duration. Meanwhile, the family had moved to Center Street on the east side near the lake. This site was located on a long hill, good for limited sleighing. It was also adequate for limited family activity, such as scrubbing, washing, and ironing operations that the female members of the family relied upon. It was necessary to deliver the washing and other output to customers, in this instance to the W. T. Kohler family on 6th Street. The convenient delivery would be by mother and Tante, usually at night, while I was for all purposes asleep upstairs. Since the downstairs facilities at Center Street were subject to spring overflow, it was deemed best by the family to restrict my user [sic]. At this interference with the family delivering right of my American user, I raised what were adequate verbal protests to the assembled neighborhood, who could not bring me succor but wanted to help. I am not sure how this eventuated except that all future Kohler deliveries were with my accompaniment. It was on one of these occasions that the Kohler sisters gave me an adequate treat of table bananas which I at once gave the American cognomen of Amerikanische Gurken. It was in such various ways that the Americanization of Martin Glaeser went on apace.

In order to develop this account without undue delay, I will skip to the next episodes that illustrate how common events disfigure and transfigure the story of our lives.

When the Milwaukee job had run its course, father returned to Sheboygan to find a new way out. I have no means of knowing how this eminent business decision was finally reached. I am sure that there was no data-processing or economico-mathematical thinking. However, there had to be some capital contribution, some supplying of fixed facilities in the form of a rented rundown shop, the notorious imported boring machine, a collection of hand tools, etc. These, together with rental equipment, were all father had to work with, in order to turn out exemplars, his son and wife, domiciled at a new nearby home, gave him the necessary company into all hours of the night. I have no means of knowing how many coffee roasters fitted to kitchen stoves were finally turned out or what they cost. This new and hopeful business in America had its marketing aspect in the form of a sales agent. A friend of my mother's, whose husband was a piano tuner by the name of Robert Werner, expressed some interest in taking orders and disposing of new ones. At this particular time, there was a disposition of American customers to buy raw materials, as from the Jewel Tea Company, who would need roasters to finish off the product. Here was an economy that looked like the American way of life.[1] His marketing area was Sheboygan and Manitowoc County.

After a reasonable period, producer and marketing agent began to have misunderstandings, which came to a head when the artistic temperament of the piano tuner collided with the same artistic but divergent understanding of the tinsmith. This being a period of direct action, the psychology of sales met the psychology of production without an intervening intermediate. Father without justification visited physical injury upon his agent. The result was a compensatory judgment by a court which the contributing tinner pronounced worthwhile. Thus ended a first commercial transaction.

Our next move resulted from a new job opportunity. As business gradually picked up so did Mr. Koepsell's job opportunity to supply facilities to the growing marine industry. This meant a removal to the south side of the city, which was close to the harbor, and marine activity, like the C. Reiss Coal Co. Our new home was at 928 Georgia Avenue, again in a region where there were swamps but closer to father's tinshop. Incidentally, this shop, next to the Sheboygan Chair Co., is still in situ.

However, new developments were continuously aborning,

It was during this period also that I found my way to a nearby parochial school called the first class of the Lutheran Dreieinigkeits Gemeinde. The importance of this school was to supply further learning of English.

The lack of success in the sale of coffee roasters induced father to become a dependent tinsmith by working as a first class tinner, joining the sheet-metal workers union and becoming an employee of the J. P. Koepsell Hardware

Company. Sheboygan, being a port city and a woodworking town, was quite a center of activity in addition to factory work and building.

Father's boss, in addition to knowing service work, was employed as a state factory inspector. This was the beginning of factory regulation under Governor Robert M. LaFollette, Sr., and was the beginning of much argumentive conversation between Mr. Keepsell as a political progressive and father as an observer of economics. With the removal of father's activities to the south side of the city, we soon occupied a series of rental properties which were cheaper than those of the north side and more in keeping with our income level. As I recall it, work at Koepsell was for 10 hours per day with additional shortened time at noon for Saturday afternoons off. Consequently, I carried lunch to my father. As long as I can remember, mother helped out the family exchequer with special activities. Tante Paula soon returned to Germany because her work there had become satisfactory. After living temporarily at three locations, we finally settled on South 12th Street, occupying the lower quarters while the owner, a barber, lived upstairs. He was an interesting person because as a barber he had time to teach me English, and he also tried to have me acquire Luxemburger which was his own native tongue. These years on the south side were very important because I began to show more independence, and participating with others, mostly children of German extraction, in games such as baseball, marbles, tag, and pump-pump pullaway. The centers of these activities were the two adjoining school grounds, one Protestant belonging to the Bethlehem Society and the other Catholic belonging to the St. Peter Claver Society. My most important companion was one Robert Werner who attended the neighboring Catholic school. His mother, likewise Catholic, had become a friend of my mother's; and they both spoke only German. The father, a piano teacher but primarily a tuner, was seldom at home and had broken with father since the late misfortune in the sale of coffee roasters,

Robert's most intrepid activity was to crawl under the table to bother me by pinching my leg when I was playing the violin with my teacher, a Mr. Schulz who would come to the house to use our table. At other times, I would go to Mr. Schulz's domicile in the upstairs room of a house which was under water but served as his home. Father's idea was to help Mr. Schulz acquire a livelihood which he needed,

Another association with the 12th Street home was the beginning of my reading of literature. Father somehow had come upon someone who sold him a German literary magazine which carried continuous stories of a semi-historical nature. The inimitable piece was the story of Mary, Queen of Scots. Of course, I fell in love with Mary and her enemies were my enemies. Without any understanding of facts, the understanding was purely romantic – but it was the beginning of an interest in history.

Before long there came a radical change. I recall hearing my parents talk about bringing grandfather and grandmother and a young cousin over from Germany because they would be able to make a better living here – grandfather as a tailor. In the meantime, father had made some social connections, the most important were with an organization known as the Deutscher Landwehrmanner Verein. The purpose of this association was to continue love for the old country and to keep up associations here. There may have been other purposes. In my father's case there were hopes he might build contacts with others who wanted German tailoring, which was always more reliable and longer lasting. My mother, who was also a tailor of no mean order, ventured the opinion that it would be impractical. But father brushed this aside and the second contingent arrived. It was not too expensive for father whose collaboration with nature was well known; it was as a new idea. He would rent a small farm on the outskirts of Sheboygan so that this combination of cottage labor, small scale farming by women, with mechanical labor would provide the economic answer. I was only interested in the fact that a change was coming, and that I would have three cows to watch and an orchard to tend. I do not want to recount the disappointments that recounted my own and others' disappointments. Had it not been for my father's abilities to expand his efforts to meet his needs, we would have suffered economic shipwreck. The only thing that I salvaged from the wreck were memories. My job was to pasture three cows who always managed to escape from the three acres where they were confined. But what a joy to read "Robinson Crusoe" and follow the descriptions of "Two Years Before the Mast."

With the fiasco of joint agriculture and industry before us, father tried a new tack. Instead of paying rent, why not own an equity? Again we moved to the east side near Lake Michigan where father bought a modest lot and constructed a house. The remainder of the activity was that we were in the city once more, near the Lake and Koepsell's tinshop. I lose the connection of grandfather and grandmother at this point. After splitting hundreds of cords of wood in a new location, this granddaughter, Johanna by name, daughter of a nailmaker spent her time at the chopping block.

If I may prognosticate as to the future of a life which has only just begun, our removal to South 8th Street was what separates America from Europe. Our new location in the end required us to separate from the German Lutheran Church and to reunite with Sheboygan in the north side.

The biggest step was the confirmation from the Lutheran Church. The difficulty here was a matter of age since confirmation was at age 13. I was only 12 so I was required to spend one more year at school. The teacher in the final year was named Piel and his attitude was to help me overcome the disadvantage of this final year in his charge. This was done by giving me extra work to do, and especially by making me the first librarian of the parochial school. Although most of the books were of a religious nature from the St. Louis School Conservatory, the work was bearable. By confirming me at the church in March, it was possible for me to enter the public school and try finishing work in the 7th grade by June. One more year in the 8th grade finished my ordinary grade school activity and made me available for high school.

The four years of high school from 1902 to 1906 were most helpful. Useless outside activity had passed. Besides the usual domestic employment, a new form of employment appeared on the scene. On account of its geographic background, Sheboygan, in fact, the entire Lakeshore, was important to the development of pea, bean, and corn production. In fact, it still is but is now less combined with the labor of men, children, and women. Technology and automatic machinery have changed all this. It was the shifting from occupations in field work to the employment in industrial processes in factories. This was a great help in providing remunerative employment. In addition to seasonal factory employment in pea factories, there was seasonal factory employment in furniture and other plants. Much economic help was thus provided which furnished a democratic contact between those who needed to work and those who were more advantageously situated. Gradually I found myself so situated that I could look forward to a better type of employment. The high school called attention to variations in capacities. The most important shift came at the end of the high school years. Those who could afford to do so went to technical or special schools or they dropped back to whatever employment could be found. The years of combination and entertainment in athletics, dramatics, and play-acting were soon no longer available. This was the period in life when all Americans had to find their place in the new economy.

My own view of life in Sheboygan was subject to new changes. Once more the harbinger of agriculture was to create a demanding problem. Father was at an age when employment as a tinsmith was satisfactory provided he wanted it as a permanent activity. Another relative, the husband of my mother's sister, was very anxious to leave factory employment to become a farmer. He lived in Sheboygan Falls and worked in a plant that finished chairs and other furniture. This suggestion induced my father to try his luck once more at farming. He purchased a 20-acre farm on the northwest side of Sheboygan which had adequate facilities. So far as I was concerned, my ideas had taken an intellectual direction. The senior instructor in the parochial school and the minister had urged my father to permit me to become a preacher or teacher. While quite satisfactory to mother, father's ideas were of an entirely different order. This was especially true when the President of Ripon College appeared at our rather modest home on 8th Street to try to convince my father to accept some scholarship assistance for me. I can still see the down-to-earth tinsmith arguing his receptacle of theoretical information against the professor. I was partly embarrassed and partly proud.

Now to return to high school – one of the principal teachers was a Mr. Winkenwerder who taught botany and physical geography. His great interest was forestry and much of his interest in this subject crept into his other subjects. Father and grandfather, as old woodsmen, were also interested along these lines.

Another teacher who greatly expanded my interest was Mr. Howe in history. His specialty was American and ancient history which he managed to combine with public speaking and debating. Because these latter subjects were in the nature of extracurricular activities, they carried a great deal of weight. As a teacher, Mr. Howe had trouble with his speech and for this reason had some difficulty with his students. He was interested, however, in having us show an understanding of the subjects beyond the elocutional point of view.

Debating and public speaking were the special interests of Mr. Howe and Miss Shepherd. The latter tried to combine an interest in politics with public speaking. Because I was president of the senior class, Miss Shepherd suggested that I undertake to represent the class in a declamatory contest which was planned to be terminated at the University. I did not have much hope of going very far. Miss Shepherd insisted, however, that I was made for the piece, "Monsieur Beaucaire" by Boothe Tarkington. After stumbling through an early attempt, I finished off at Music Hall at Madison, Wisconsin, much to the surprise of the Sheboygan contingent. If there had ever been any doubt about my matriculating at the University, this doubt was soon lifted. It included even the promise of a meal job where the Sheboygan students were well represented.

I must not omit to mention my own obligation to assist Herbert Kohler, who was a member of my high school debating club. Interest in debating terminated in an annual joint debate between two groups, the Ciceronians and the Spooners. It was the order of the day. The final selection of interschool debating teams – Sheboygan vs. Manitowoc and Fend du Lac – constituted the final windup. The membership at the time constituted Albeit Axley, the writer, George Eberle, and Gustave Buchen, and Marie Schufflebotham. Mr, Buchen continued his interest in intercollegiate forensics, terminating as state senator and as the author of a comprehensive history of Sheboygan County. Even outside the circle of the high school, there were interested sponsors of debating as A. C. Hahn, General Manager of the Phoenix Chair Co. Yet, no one will ever forget the great dame of them all, in theatricals, in elocution, in debating, in outside training of students, Marie Kohler, whose loss to Sheboygan and accession to the village of Kohler was irremediable.

In the year 1910 the interests of a group of Sheboygan alumni seem to have concentrated about the idea of a vagabond trip to Germany and England. One of the reasons was the appearance of a small number of travel books which made much of the peaceful condition of the international world and the economy of travel throughout.

Also during the summer of 1909 a group of more intimate friends had taken up a proposal by Mathew Schiller, a tobacco salesman in Sheboygan, to lead a group of vacationers on a sail boat journey around Lake Winnebago. This induced Mr. Schiller to join the Sheboygan contingent and to enroll in the University as a freshman in the course in industrial education. He hoped to make use of his experience as a boat builder, piano player and piano tuner. In addition he earned his keep as a barber and doing odd jobs. What sustained him was the promise of his trip to Europe, partly compensated by these friends. The journey, however brief, was successfully concluded and was the cause of considerable envy.

The successful conclusion of a series of job engagements during the school year 1911 enabled me to place the objective of this trip to Europe as among the attainable achievements available to me. The most important promise of achieving this result was the half-time post as statistical clerk of the Railroad Commission, serving under Cecil Schreiber at $30 per month. A similarly, if irregularly, compensated post was provided by an accounting clerkship under the University Extension Division correcting business administration assignments. These were further supplemented by the time-honored opportunity to work for my meals by doing odd jobs at Phi Alpha Delta sorority. What seemed to cap a climax was the firm promise of a collegiate friend from Chicago of a cattleboat job from Montreal to Liverpool. As for the rest of the trip, we felt the Lord would provide. With the accompanying gift of suitable corduroy suits of clothes by another colleague, Al Schorting of the Milwaukee Reiss-Friedman Co., we were ready for June departure. We in this case constituted as the principals – Schiller with lots of travel experience, the writer, and a graduate colleague, Louis Augsburger from Sheboygan, who was a candidate for the Ph.D. in Chemistry at Bonn University. In the end, our opinion was that Louis wanted to go along for the ride.

Our trip to Montreal was uneventful. However, we looked up the ship, the Magentic, at its berth, and that is where things began to happen. The general filth and disorganization about the ship induced Louis to give up his cattle boat job opportunity and to move up to steerage, which was priced at $30. Second class passage which was available to him would have been considerably higher. After the necessary compromise we boarded the ship as pay-steerage passengers.

There was no comparing my first crossing of the Atlantic with this one, although the ocean route was about the same. This ship was all steam. Its first few days of sailing on the St. Lawrence were landlocked and we saw the battlements of Quebec. Once at sea we saw icebergs which were a menace to shipping. The trip to Liverpool was short and not uncomfortable. I remember seeing nothing of the cattle.

We were alarmed by hearing that all major English ports were strikebound and that our further progress toward Hamburg was uncertain. Augsburger left us here because he wanted to see London.

At first we decided to see a little of Liverpool. It was all so new. By accident we wandered into a depressed cemetery. This interested us because old gravestones nearly always provide historical interest. This was no exception, for here was the marker of the man who invented the iron rail. While we were lingering over this site, some person., presumably a native, accosted us as foreigners and asked to show us about. Showing us about included a visit to the trade union strike center which in the end was cleared by the military.

The rail trip across the island from Liverpool via Manchester to Grimsby on the south side of the Humber River was uneventful, but the port was over-crowded. We were told we would have difficulty finding transport to Hamburg. The hold of the ship was jammed, so were the approaches to the decks. Mr. Schiller, recognizing the confusion, threw his suitcase over the deck and hurried downstairs. I followed him. Since we had not eaten since leaving Liverpool, the problem of hunger appeared. It was late afternoon when we boarded the ship. The regular passengers were provided with two meal tickets, which we did not have and could not get without exposure. My mother had packed a substantial lunch consisting of homemade summer sausage, and we ate this with the smell and taste of herring juice leaking out of the boxes stored where we were hiding. To escape the smell we hid on deck and slept above the firehole of the engine room. The crossing was reasonable, but it was a bad night. However, by 9 o'clock we sailed up the Elbe River and landed at Hamburg Customs with our passports.

Our next rail stretch was a fourth class trip through the wooded area of the Mark Brandenburg with Berlin as our destination. I do not recall many of the incidents of our stay in that metropolis. It was, of course, punctured by German militarism. Still the Kaiser and his tradition had its opposition. I recall the Socialist tradition in their political paper, the Kladeradatsch, which made fun of Court festivities, to wit: Wie bei Kaisersgetanzt wird.

We were at a great disadvantage in that we had no private transportation at our disposal. We had to depend upon public agencies, which fortunately were adequate in Berlin, including rapid transit. This enabled us to visit outlying places like Charlottenburg and Potsdam, the latter made sacred as the home of Frederick the Great. Some idea of the outlook and understanding that Prussian militarism was able to shed appears from the statement attributed to Napoleon. Upon the occasion of his visit to Potsdam early in the 19th century, some of his Generals and Marshals behaved less than honorably. He rebuked them: Gentlemen, if the man before whose bier we are standing were alive today, we would not be here.

It was not possible for Mr. Schiller and me to visit all places of artistic, historic, and academic interest. We did visit the Tiergarten, which has now been reduced in area to a few acres but at that time was an appendage to the Bradenburger Thor and the Siegesallee. These gave indications of the nationalistic splendor of the state.

While walking along the main street, Unter den Linden, there suddenly appeared from the eastern sector a cavalcade of fast-moving automobiles with piercing horns and flaunting flags. It was the Kaiser and his procession daring to show himself to his people. It was this unusual display that made me realize that we were not in Europe to see Royalty but to visit relatives who were like ourselves. So next morning found us aboard the express train bound for the southwestern section of the country – the former province of Silesia. The first relative we undertook to visit was my mother's oldest sister at Peterswaldau, who continued to be our most frequent correspondent and whose son, Ludwig Poltmann, wrote most frequently to my father. He did not like father's Americanism, but their mutual criticism was well meant and its content was definitely meant to be on as scientific basis as possible. For instance, father had graduated from a so-called handwerker-fortbildungschule, while Ludwig was a Kaufmann. There was some use of mutual intelligence and common sense, although an appreciation of real institutional variation was definitely lacking.

Our next visit was to Tepliwodau, my birthplace. There we visited father's friends from the Schenke and particularly the salesman and repairman of automobiles who was his special friend. We were asked to stay overnight at the Schenke, which very nearly turned out to be dangerous for Mat Schiller. Next day the auto salesman accompanied us to Breslau to attend a Kommers of the Breslau alumni. He called our attention to people of prominence, especially the Kronprinz, as representative of the Kaiser. Many people of prominence had shaved off their beards to reveal their "Schmisse" or marks of dueling. Upon our return we once more noticed points of interest at Tepliwodau, and especially the profusion of ducks and geese belonging to the villagers.

With a quick auto trip to Stadt and Festung Silberberg, at the border between Germany and Bohemia, which was made famous as the beautiful prison of Frederick the Great before he became king, we completed this circuit.

At my grandfather's request we next visited the famous city of Goldberg. As its name indicates, it was a well known gold mining town – one of the very few in Europe. Its history extended back to the year 1000 with the separate events chronicled by monks. Mining activity in this city was important to the German Emperor and the Dukes of Poland and Liegnitz. There were schools of higher learning, such as a famous Latin school set up by the Holy Roman Emperor. General Wallenstein was an alumnus of the school, who during the 30 years war failed to protect the city.

It was a peculiar circumstance that a short-line railroad provided occasional transport to the city, whose passenger and freight station was at the bottom of a hill; nevertheless, governmental, commercial and social centers were at the top of a series of stairways. Grandfather's shop and residence had been on this Ringstrasse. Except for the age of Goldberg, there were other cities important to my father. Flinzberg was father's birthplace and the home of the Glaeser clan. It also was the beginning of the ascent to the Schneekoppe, the highest mountain of Germany. Fauer was the city where father served in the army and which Tante Paula called home. Most of these locations were west of the Oder River and east of the wuthende (raging) Neisse. Hence, all of this area is now claimed as a part of Poland and looked upon as a German irredentist area.

Our next stop was Dresden where my mother's youngest sister, Ernestine, was then living.

In order to expand and develop our understanding of German culture and institutions we planned our next stop to be Dresden, the capital city of the Kingdom of Saxony. This had a close connection with our family as it was the home of my mother's youngest sister, Ernestine. Her deceased husband and she had been employed there as textile workers, an industry for which Dresden was famous. At one time it had been a noted art center compared to the Italian city of Florence. The King of Saxony (August der Starke) had at one time been the King of Poland and was responsible for much of its industrial development. In the concourse of its German principalities it was known as das rothe Konigreich (the red kingdom). My aunt was anxious to have us visit her and meet her present employer, a well-to-do business man. After an ever welcome dinner at his home, he took us to the Imperial Gallery with its famous paintings and portrait gallery. In the evening a special treat was arranged and we visited a scientific museum devoted to biological findings – the most important denominated Der Mensch. Among many unique findings was a large series of human lungs beginning with examples of mountaineers and ending with the lungs of coal miners.

In planning our further progress we realized that most of the financial help had come from my relatives and friends and that we missed the $30 consumed in the fare of the Magentic. When I told my aunt that we were running short, she presented me with funds that she had saved through social security. Of course, I refused and made out that we could somehow find our way. It was here that Mr. Schiller and I had our only misunderstanding; because he came up with $25 which he had saved in order to buy a German telescope which he had been unable to buy last time. This really was the beginning of saving and no fooling.

Our next stop was to be Frankfurt, the sacred legacy of Goethe and Schiller. We had been advised always to stay at the restaurants frequented by cab drivers and truckers. Not only were meals good and cheap, but some were paid for by gratuities in return for singing American songs and telling tales. Here is where Mat excelled. Not only did he have a mellifluous tenor, but he had no hesitancy in using it.

Our last opportunity to get acquainted with German small-scale industry was when we visited Goslar in the Harz Gebirge. This came about as a result of Professor Ernst Feise of the German Department, who urged us to visit a German Professor who lived at Goslar, but whose circle of activity extended throughout the Province. He functioned as a governmental economic supervisor of the investment of insurance companies. We were promised an extended automobile tour throughout this scenic province, which was to last for about a week. Professor Feise was a relative of this German Professor, according to the German institution of Wahlerwandshaft (voluntary relationship). Arriving at Goslar about noon we decided to have dinner before calling upon our party who lived apart in what the Germans called a Schloss. In celebration of the occasion we each bought an expensive Dutch cigar.

After announcing our arrival by ringing a heavy gong, we were finally admitted to a reception room where a lady sat propped up in many protective bandages. She apologized for her present appearance and said she was substituting for her husband who had suffered a serious automobile accident and was still in the hospital. However, she wanted to express the family's disappointment that it was unable to meet their brother Ernst Feise's requests.

Our statements of regret were certainly more than honest. After expressing hope that we would be able to reach Cologne before nightfall, we hurried to the railroad station, more than conscious that emergencies made up more than our share of luck. The military commanders at Osnabruch and Munster during the Thirty Years War were not the only losers.

In general our ability to understand the magnificent contributions made by the Cologne cathedral to the development of the commerce of the Rhine was apparent. The tremendous accumulation of precious metals as a basis for commerce and trade made us realize that the focal point was not only Cologne as the head of tourist traffic, but this also extended beyond the Lower Rhine into Belgium and Holland and the points of trans-shipment to England and France. Why was this not an opportunity, like our earlier transport across the North Sea? If the opportunity to stow away to Hamburg via the Elbe was successful, why not use the goods traffic on the Lower Rhine boats in similar fashion. Any attempt to hide below would be a great lift along the way we were going. However, we failed to take account of the watchfulness of a Dutch steward who guessed our plan. However, a new variant of fortune appeared. An Englishman and his wife en route to Gloucester were similarly interested in an all-water trip home. He took note of our predicament and offered to pay our fare as from an Englishman to a shipwrecked American. Not only did this solve our immediate problem, but it put me in mind of an academic assist.

As a student of Ernst Meyer, one time consular officer at Chemnitz, an offhand remark made by Professor Meyer that among other duties it was a consul's duty to assist in sending home shipwrecked sailors, we decided to try out his theory. Would this not then be an answer to our major problem? With this problem on the way to solution, we were certain that the American consul at Rotterdam held the key to our troubles. To make sure that we would have transport privileges back home, in Berlin we had paid down 10% of our return fare. We were relying upon Providence that the remaining 90% would somehow be made available. Repeating our previous experience at Goslar of purchasing expensive cigars, we purchased two theatre tickets which almost landed us in a Dutch canal. Some hay floating on the canal deceived us and we thought it was terra firma. A special theatre performance was given for the young queen, Juliana, so we got some idea of royalty with the Dutch enthusiasm.

That night we slept hopefully. All problems seemed to be solved with the picture of Bartholdy's Liberty floating in my mind. Next morning we called on the American Consul at Rotterdam and found that he was barely able to speak English. He refused to classify us as shipwrecked sailors. Instead, we were treated as indigent tourists who had been in Europe for a good time. I don't recall how many times we called upon the Consul nor what our arguments were. Finally he said. "Boys, I believe you are in earnest." I recall Oscar Wilde's play in which I performed "The Importance of Being Earnest." One never knows what academic lore will deliver you.

Next day we boarded the S. S. Rotterdam as steerage passengers. Mat and I let our joy be somewhat unrefined, to the dismay of two other steerage passengers, a Turk and a Bulgarian, the former unable to understand English while the latter was reading Faust in German.

NOTE

1. Alfred Marshall explained "the superiority of the American inventive faculty over the English" on acculturized consumer habits: "Chiefly by the difference in the habits of the consumers: English people are contented with the things to which they are accustomed. Americans are always on the lookout to see if they cannot find something better suited to their purposes than they have already. The demand for ingenuity calls forth the supply" (John C. Whittaker, *The Correspondence of Alfred Marshall, Economist*. Three volumes. New York: Cambridge University Press, 1996. Marshall to Rebecca Marshall, June 12, 1875, entry of June 10, 1875, letter 22, vol. 1, p. 49; on the socialization of immigrants' children, see Marshall to Rebecca Marshall, September 5, 1875, letter 30, vol. 1, p. 77)). Of course, Martin Glaeser's father was an immigrant from Germany with certain skills which he was ready, able, and willing to put to use; he thereby both fit in well with his new culture and helped reinforce it.

TOWARD A BIBLIOGRAPHY OF EDGEWORTH'S WRITINGS*

Alberto Baccini

After a long period of oblivion, interrupted only by the essays by Stephen Stigler (1978) and Peter Newman (1987), and by the book of John Creedy (1986), there seems to be a renewed interest for the works of Francis Ysidro Edgeworth.[1] The appearence at a short interval of two re-editions, though very different in aim, of Edgeworth's paper on probability and statistics – the book edited by Philip Mirowski in 1994, and the three volume set edited by Charles R. McCann in 1996 – testify this new interest.[2] However, scholars do not yet have a complete bibliography of Edgeworth's works. The aim of the following pages is to try and fill this gap. The reason for the lack of a complete bibliography of Edgeworth's writings is probably that he never wrote treatises. The difficulty of Edgeworth's texts, traditionally noted by scholars, and their dispersion, delayed or impeded their appreciation – as noted by Bowley (1928, p. 3) – as well as their complete collation.

If we do not consider the "List of writings upon cognate subjects by the same author" which Edgeworth inserted at the end of *Metretike* in 1887, we have only three attempts to compile a bibliography of Edgeworth's writings. The first attempt – limited to the works on probability and statistics – was the "annotated bibliography" in Arthur Lyon Bowley's *Edgeworth's Contributions to Mathematical Statistics,* published in 1928. This bibliography did not include the papers in economics reprinted in the *Papers Related to Political Economy*, edited by Edgeworth himself, and published by the Royal Economic Society in 1925.

Documents on Modern History of Economic Thought, Volume 21-C, pages 271–301.
© 2003 Published by Elsevier Science Ltd.
ISBN: 0-7623-0998-9

A second attempt was made at the beginning of the Fifties by Harry Johnson who co-ordinated a research team at Cambridge. The outcome of this research remained, however, unpublished. A mimeographical version of this was circulated from Chicago (Stigler, 1978, p. 318). As noted by Newman this bibliography is "the most useful" but also "rather inaccurate and incomplete" (1990, p. 132).

The third attempt has been conducted by Mirowski, who published "An Incomplete Bibliography of Edgeworth's Writings" as an appendix to Mirowski (1994). His main sources were Bowley (1928), and Stigler (1978), and his bibliography contains some items not previously noted: in particular, Mirowski attributed to Edgeworth some reviews, the authorship of which is very difficult to establish.[3]

The lack of a nearly complete published bibliography has given rise to very different *estimates* of the number of Edgeworth's writings. Johnson's bibliography lists 173 articles and books; 149 book reviews, and 131 entries for the *Palgrave's*; Clifford Hildreth (1968, p. 507) wrote of "7 small books and numerous journals articles" without adventuring himself in a precise estimate; Stigler (1978, p. 311) mentioned 200 book reviews, over 130 articles in the *Palgrave's*, and three short biographies in the *Dictionary of National Biography*; Newman (1987, p. 84) counted four books, 172 articles, 173 book reviews, 136 entries in the *Palgrave's*; Mirowski (1994) included in his "incomplete bibliography" a total of 271 entries – counting multi-part articles as several ones – including only 118 book reviews; according to McCann (1996, p. xiv), from 1883 to 1926 Edgeworth published over 170 articles, including review articles.

The bibliography presented here includes 4 books; 198 articles – if we count multipart articles separately we arrive at 233, – 204 reviews and 139 entries in the *Palgrave's*. The items previously unpublished are 9 articles and 22 book reviews; the most part for the period 1882–1885.

Following Johnson's, my bibliography is ordered chronologically and organized in four sections: section A contains Edgeworth's books;[4] section B includes articles; section C book reviews, while section D comprises the entries of the *Palgrave's Dictionary of Political Economy*. This bibliography – more "conservatively" than Newman's counting (1990, p. 131) – adopts Johnson's classification for articles and book reviews.[5] It is then possible to consider Johnson's bibliography as a subset of ours: the items present in Johnson's bibliography are those which do not have annotations.

As in Johnson's bibliography books and articles are listed in order of publication. Multipart articles are listed as only one entry if the parts were published in the same year; in two or more entries if the parts were published in two or more years. Square brackets contain the indication of the reprint in the three volumes of *Papers Relating to Political Economy* (1925);[6] roman numbers indicate the volume. If the title of the original article does not match the title of the reprint, we also include the title of the reprint.

Book reviews are listed in order of publication; for brevity, they are listed with the indication of the author and the title of the work

The following typographical symbols and abbreviations indicate:

© items not present in Johnson's bibliography;
\# items reprinted in McCann;
§ items reprinted in Mirowski;
@ items listed here for the first time to our knowledge.

AC: *Academy*;
BAR: *Reports of the British Association for the Advancement of the Science*;
EJ: *Economic Journal*;
GE: *Giornale degli economisti e rivista di statistica*;
HERM: *Hermathena. A series of papers on Literature, Science and Philosophy by Members of Trinity College, Dublin*;
JED: *Journal of Education*;
JRSS: *Journal of the Royal Statistical Society*;
PHM: *London, Edinburgh and Dublin Philosophical Magazine and Journal of Science*;
PRPE: Francis Ysidro Edgeworth (1925). *Papers relating to Political Economy.* London: MacMillan and Co.
QJE: *Quarterly Journal of Economics.*

Section A. Books

1. *New and Old Methods of Ethics, or "Physical Ethics" and "Methods of Ethics"*, Oxford & London, James Parker and Co., 1877.
2. *Mathematical Psychics. An Essay on the Application of Mathematics to the Moral Sciences,* London, Kegan Paul & Co, 1881 (photographic reprint London, LSE, Series of reprints of scarce tracts in Economic and Political Science, 1932; also reprinted by Augustus M. Kelley, New York, 1967).[7]
3. *Metretike or the Method of Measuring Probabilities and Utilities*, London, Temple Company, 1887.§
4. *Papers Relating to Political Economy*, London, MacMillan and Co., 1925 (New York, B. Franklin, 1970).

Section B. Articles

1. Mr. Matthew Arnold on Butler's doctrine of self-love, *Mind*, 1876, *I*, 570–571.

2. The hedonical calculus. *Mind*, 1879, *IV*: 394–408.
3. A summary of *Mathematical Psychics. Mind*, 1881, *VI*, 293.@
4. Obituary of William Stanley Jevons. *AC*, *538*, (August 26)1882, 151–152.@
5. Mr. Leslie Stephen on utilitarianism. *Mind*, 1882, *VII*, 446–447.§
6. The abstract theory of rent (Précis of paper given before section F). *BAR*, 1882, 642.
7. The law of Error. *PHM*, 1883, *XVI*, 300–309.#
8. The method of least squares. *PHM*, 1883, *XVI*, 360–375.#§
9. The physical basis of probability. *PHM*, 1883, *XVI*, 433–435.#§
10. On the method of ascertaining a change in the value of gold. *JRSS*, 1883, *XLVI*, 714–718.#§
11. The rationale of exchange. *JRSS*, 1884, *XLVII*, 164–166.§
12. The philosophy of chance. *Mind*, 1884, *IX*, 223–235.#§
13. On the reduction of observations. *PHM*, 1884, *XVII*, 135–141.#
14. *A priori* probabilities. *PHM*, 1884, *XVIII*, 204–210.#§
15. Chance and law. *HERM*, 1884, *X*, 154–163.#
16. Baker Thomas. In: *Dictionary of National Biography* (Vol. 1, Part 3, 1908, p. 937). London, Smith, Elder and Co., 1885.©
17. Balam Richard. In: *Dictionary of National Biography* (Vol. 1, Part 3, 1908, p. 944). London, Smith, Elder and Co., 1885.©
18. Baxter Thomas. In: *Dictionary of National Biography* (Vol. 1, Part 3 (1908, p. 1352). London, Smith, Elder and Co., 1885.©
19. On the methods of ascertaining variations in the rates of birth, death and marriage (Précis of paper given before section F). *BAR*, 1885, 1165–1166.#
20. The calculus of probabilities applied to psychical research. *Proceedings of the Society for Psychical Research*, 1885, III: 190–199.#§
21. On methods of ascertaining variations in the rate of births, deaths and marriages. *JRSS*, 1885, *XLVIII*, 628–652.
22. Methods of statistics. *JRSS*, 1885, (Jubilee volume), 181–217.#
23. The calculus of probabilities applied to psychical research. II. *Proceedings of the Society for Psychical Research*, 1886, *IV*: 189–208.#
24. The law of error and elimination of chance. *PHM*, 1886, *XXI*, 308–324.#
25. On the determination of the modulus of errors. *PHM*, 1886, *XXI*, 500–507.#
26. Problems in probabilities. *PHM*, 1886, *XXII*, 371–384.#
27. Progressive means. *JRSS*, 1886, *XLIX*, 469–475.#
28. The mathematical method of statistics. *JRSS*, 1886, *XLIX*, 649–654.#
29. The mathematical theory of banking, *BAR*, 1886, 777–779.#§
30. The element of chance in examination (Précis of paper given before Section F). *BAR*, 1886, 920.

31. Observations and statistics: an essay on the theory of errors of observation and the first principles of statistics (read May 25th, 1885). *Transactions of the Cambridge Philosophical Society*, 1887, *XIV*(II), 138–169.[8]#

32. Observation and statistics corrigendum. *Proceedings of the Cambridge Philosophical Society*, 1887, *6*, 101–102.

33. On discordant observations. *PHM*, 1887, *XXIII*, 364–375.#

34. A new method of reducing observations relating to several quantities. Letter calling attention to article in HERM. *PHM*, 1887, *XXIV*, 222–223.

35. The choice of means. *PHM*, 1887, *XXIV*, 268–271.#

36. Empirical proof of the law of error. *PHM*, 1887, *XXIV*: 330–342.#§

37. The method of measuring probability and utility. *Mind,* 1887, *XII*, 484–485.[9]

38. Report of the committee appointed for the purpose of investigating the best methods of ascertaining and measuring variations in the value of the monetary standard (drawn up by the secretary). *BAR*, 1887, 247–254 [reprinted with 39 and 51 in *PRPE: Measurement of change in the value of money*, *I*, 195–297].

39. Memorandum by the secretary. *BAR*, 1887, 254–301 [reprinted with 38 and 51 in *PRPE: Measurement of change in the value of money*, *I*, 195–297].

40. The law of error. *Nature*, 1887, 482–483.[10]#§

41. On observations relating to several quantities. *HERM*, 1888, *XIII*(6), 279–285.#

42. Memorandum by the secretary, Prof. F. Y. Edgeworth, on the accuracy of the proposed calculation of index numbers. Appendix to the second report of the Committee appointed for the purpose of investigating the best method of ascertaining and measuring variations in the value of the monetary standard. *BAR*, 1888, 188–219 [reprinted with 43 and 44 in *PRPE: Test of accurate measurement*, *I*, 298–343].

43. Report of the Commitee appointed for the purpose of inquiring and reporting as to the Statistical data available for determining the amount of Precious Metals in use as money in the principal countries, the chief forms in which the money is employed, and the amount annually used in the arts (drawn up by the secretary). *BAR*, 1888: 219–224 [reprinted with 42 and 44 in *PRPE: Test of accurate measurement*, *I*: 298–343].

44. Memorandum by the secretary on Jevons' method of ascertaining the number of coins in circulation. *BAR*, 1888, 224–232 [reprinted with 42 and 43 in *PRPE: Test of accurate measurement*, *I*: 298–343].

45. The mathematical theory of banking. *JRSS*, 1888, *LI*, 113–127.

46. Some new methods of measuring variation in general prices. *JRSS*, 1888, *LI*, 346–368.#§

47. The statistics of examinations. *JRSS*, 1888, *LI*, 599–635.#

48. On a new method of reducing observations relating to several quantities. *PHM*, 1888, *XXV*, 184–191.#

49. The value of authority tested by experiment. *Mind*, 1888, *XIII*, 146–148.

50. On the statistics of examination (Précis of paper given before section F). *BAR*, 1888, 763.[11]

51. Third report of the Committee appointed for the purpose of investigating the best method of ascertaining and measuring variations in the value of the monetary standard. Memorandum by the Secretary, BAR, 1889: 133–164. [reprinted with 38 and 39 in *PRPE: Measurement of change in the value of money*, *I*, 195–297]

52. Address to the economic science and statistics section of the British Association, September 12th. Application of mathematics to political economy. *BAR*, 1889, 671–696.[12]

53. Opening address in section F at the British Association – Points at which mathematical reasoning is applicable to political economy. *Nature*, 1889, *40*, 496–509.[13] [*PRPE*, *II*, 273–310.]#

54. On the application of mathematics to political economy (presidential address to Section F of the British Association in 1889). *JRSS*, 1889, *LII*, 538–576.[14]

55. Remarks on the experiments [of Cattell, J. McK., Bryant, S., Mental association investigated by experiment. *Mind*, 1889, *XIV*, 230–244]. *Mind*, 1889, *XIV*, 245–246.

56. Report to the British Association on the variation in the value of Monetary standards. *Nature*, 1889, *40*, 553.[15]

57. Appreciation of gold. *QJE*, 1889, *3*, 151–169.

58. The elements of chance in examinations (Précis of a Paper given before Section F). *BAR*, 1890, 920.@

59. The uncertainty of examinations. *JED*, *12*, (February 1), 1890, 95–96.©#

60. The element of chance in competitive examinations. *JED*, (12 April), 1890, 203.©#

61. The element of chance in competitive examinations. *JRSS*, 1890, *LIII*, 460–475, 644–663.#

62. Problems in probabilities. No. 2: competitive examinations. *PHM*, 1890, *XXX*, 171–188.#

63. An introductory lecture on political economy. *EJ*, *1*, 1891, 625–634 [*PRPE*: *The objects and methods of political economy*, *I*, 3–12].

64. The British Economic Association. *EJ*, *1*, 1891, 1–2.

65. La théorie coût de l'offre et de la demande et le cout de production. *Revue d'Economie Politique*, 1891, *V*, 10–28.

66. Osservazioni sulla teoria matematica dell'economia politica con riguardo speciale ai principi di economia di Alfredo Marshall. *GE*, 1891, *I*, 233–245.

67. Ancora a proposito della teoria del baratto. *GE*, 1891, *II*, 316–318 [English Translation in *PRPE, On the determinates of economic equilibrium, II,* 313–319].[16]
68. Recent attempts to evaluate the amount of coin circulating in a country. *EJ*, 1892, *II*, 162–169 [*PRPE: Evaluation of metallic currency, I*, 406–415].
69. Correlated averages. *PHM*, 1892, *XXXIV*, 190–204.
70. The law of error and correlated averages. *PHM*, 1892, *XXXIV*, 429–438, 518–526.#
71. Statistical correlation between social phenomena. *JRSS*, 1893, *LVI*, 670–675.#§
72. A new method of treating correlated averages. *PHM*, 1893, *XXXV*, 63–64.#
73. Exercises in the calculation of errors. *PHM*, 1893, *XXXVI*, 98–111.#
74. Note on the calculation of correlation between organs. *PHM*, 1893, *XXXVI*, 350–351.
75. On statistical correlation between social phenomena (Précis of a paper given before Section F). *BAR*, 1893, 852–853.
76. Theory of international values. *EJ*, 1894, *4*, 35–50, 424–443, 606–638 [*PRPE: The pure theory of international values, II*, 3–60].
77. Professor J. S. Nicholson on «Consumer rent», *EJ*, 1894, *4*, 151–158.
78. (with Higgs, H.), One word more on the ultimate standard of value. *EJ*, 1894, *4*, 724.
79. The measurement of utility by money. *EJ*, 1894, *4*, 347–348.
80. Asymmetrical correlation between social phenomena. *JRSS*, 1894, *LVII*, 563–568.#
81. Professor Böhm-Bawerk on the ultimate standard of value. *EJ*, 1894, *4*, 518–521, 724–725 [*PRPE, III*, 59–64].
82. Recent writings on index numbers. *EJ*, 1894, *4*, 158–165 [*PRPE, Variorum notes on index numbers, I*, 344–350].
83. The asymmetrical probability curve. *Proceedings of the Royal Society,* 1894, *LVI*, 271–272.#
84. On the statistics of wasps, BAR, 1895: 729–730.@
85. On some recent contributions to the theory of statistics, JRSS, 1895, LVIII: 506–515. #
86. Pierson on scarcity of gold. *EJ*, 1895, *5*, 109–112 [*PRPE, I*, 351–355].
87. The stationary state in Japan. *EJ*, 1895, *5*, 480–481.
88. Thoughts on monetary reform. *EJ*, 1895, *5*, 434–451 [*PRPE, Questions connected with bimetallism, I*, 421–442].
89. Bemerkungen über die Kritik meiner *Methoden der Statistik* von Dr. V. Bortkewitsch. *Jahrbücher für Nationalökonomie und Statistick*, 1896, *XI*, 274–277.[17]

90. Statistics of unprogressive communities. *JRSS*, 1896, *LIX*, 356–386.

91. Supplementary notes on statistics. *JRSS*, 1896, *LIX*, 529–539.§

92. On the statistics of wasps (Précis of a paper given before Section D). *BAR*, 1896, 836.[18]@

93. A defence of index number. *EJ*, 1896, *6*, 132–142 [*PRPE, I*: 356–368].

94. The asymmetrical probability curve. *PHM*, 1896, *XLI*, 90–99.#

95. The compound law of error. *PHM*, 1896, *XLI*, 207–215.#

96. Eine erwiderung. *Jahrbücher für Nationalökonomie und Statistick*, 1896, *XII*, 838–845.

97. The statistics of bees. *BAR*, 1897: 694.@

98. La teoria pura del monopolio. *GE*, 1897, *15*, 13–31, 307–320, 405–414 [*PRPE, The pure theory of monopoly, I*, 111–142].[19]

99. (With Higgs, H.) Interview with McArthur (the inventor of the cyanide process for the extraction of gold). *EJ*, 1897, *7*, 119–122.

100. The pure theory of taxation, EJ, 1897, 7: 46–70, 226–238, 550–571. [PRPE, II: 63–125].

101. La curva delle entrate e la curva di probabilità. *GE*, 1897, *15*, 215–218.

102. Miscellaneous applications of the calculus of probabilities. *JRSS*, 1897, *LX*, 681–698, 119–131, 534–542.#

103. The mathematical representation of statistics (Précis of a paper given before Section A). *BAR*, 1898, 791.[20]@

104. Miscellaneous applications of the calculus of probabilities. *JRSS*, 1898, *LXI*, 119–131, 534–542.#

105. On the representation of statistics by mathematical formulae (Part I). *JRSS*, 1898, *LXI*, 670–700.#

106. Professor Graziani on the mathematical theory of monopoly. *EJ*, 1898, *8*, 234–239 [*PRPE, III*, 89–95].[21]

107. On the representation of statistics by mathematical formulae (Parts II, III and IV). *JRSS*, 1899, *LXII*, 125–140; 373–385; 534–555.#

108. Professor Seligman on the mathematical method in political economy. *EJ*, 1899, *9*, 286–315 [*PRPE: Professor Seligman on the theory of monopoly, I*, 143–171].

109. On a point in the theory of international trade, EJ, 1899, 9: 125–128.

110. Answers to questions put by local taxation commission. *Royal Commission on Local Taxation, 9528*, 1899, 126–137 [*PRPE, II*, 126–149].

111. On the use of Galtonian and other curves to represent statistics (Précis of paper given before Section F). *BAR*, 1899, 825.

112. On the representation of statistics by mathematical formulae. (Supplement.) *JRSS*, 1900, *LXIII*, 72–81.[22]#

113. Report from the Head Commissioner paper currency, Calcutta, to the secretary of the government of India, Finance and Commerce Department. *EJ*, 1900, *X*, 109–113 [*PRPE, Defence of Mr. Harrison's calculation of the rupee circulation, I*, 416–420]

114. The incidence of urban rates. *EJ*, 1900, *X*, 172–193, 340–348, 487–517 [*PRPE, Urban rates, II*, 150–214].

115. Disputed points in the theory of international trade. *EJ*, 1901, *XI*, 582–595.

116. Mr. Walsh on the measurement of general exchange value. *EJ*, 1901, *XI*, 404–416 [*PRPE, Mr. Walsh on the measurement of exchange value, I*, 369–383].[23]

117. Law of Error, in *Encyclopaedia Britannica* (Xth ed.), 1902, *XXVIII*, 280–291.[24]

118. Theoretical considerations, pp. 325–331 in Edgeworth, F. Y., Bowley, A. L. Methods of representing statistics of wages and other groups not fulfilling the normal law of error. *JRSS*, 1902, *LXV*, 325–354.#

119. The Law of Error. *BAR*, 1903–1904, 463.©

120. The theory of distribution. *QJE*, 1904, *18*, 159–219 [*PRPE, I*, 13–60].

121. Preface. In: Ramsay MacDonald, J. (Ed.), *Women in the Printing Trades, a Sociological Study.* London, P. S. King, 1904.

122. A moot point in the theory of international trade (Précis of a paper given before Section F). *BAR*, 1904, 647.

123. The law of error. *Cambridge Philosophical Society Transactions*, 1905, *XX*(I), 36–65, 113–141.[25]#

124. The generalized law of error, or law of great numbers. *JRSS*, 1906, *LXIX*, 497–530 (discussion 531–539).#

125. Recent schemes for rating land values. *EJ*, 1906, *16*, 66–77 [*PRPE, Further considerations on urban rates, II*, 215–233].

126. The rating of urban land values. *Clare Market Review*, 1906.[26]

127. Address to students. *PRPE, II*, 227–233.[27]

128. Theory of distribution (Précis of a paper given before Section F). *BAR*, 1906, 642.

129. Statistical observations on wasps and bees. *Biometrika*, 1907, *V*(4), 365–386.

130. On the representation of statistical frequency by a series. *JRSS*, 1907, *LXX*, 102–106.#

131. John Kells Ingram (Obituary). *EJ*, 1907, *XVII*, 299–301.

132. Appreciations of mathematical theories. *EJ*, 1907, *XVII*, 221–231, 525–531 [*PRPE, Variorum theories on consumers' surplus, rent, duopoly, entrepreneurs' remuneration, II*, 320–339].

133. Appreciations of mathematical theories. *EJ*, 1908, *XVIII*, 392–403, 541–556 [*PRPE, Mr. Bickerdike's theory of incident taxes and customs duties*, *II*, 340–366].

134. On certain peculiarities of small duties on imports and exports (Précis of a paper given before section F). *BAR*, 1908, 786.[28]

135. On the probable errors of frequency constants. *JRSS*, 1908, *LXXI*, 381–397, 499–512, 651–661, 662–678.#§

136. On the probable errors of frequency constants. Addendum. *JRSS*, 1909, *LXXII*, 81–90.

137. Apprezzamenti di teorie matematiche. *GE*, 1909, *I*, 635–694.[29]

138. On the use of the differential calculus in economics to determine conditions of maximum advantage. *Scientia*, 1910, *7*, 80–103 (French translation Supplement, pp. 44–69) [*PRPE, Application of differential calculus to economics*, *II*, 367–386].

139. The subjective element in the first principles of taxation. *QJE*, 1910, *XXIV*, 459–470 [*PRPE, Minimum sacrifice vs. equal sacrifice*, *II*, 234–242].

140. Applications of probabilities to economics. *EJ*, 1910, *20*, 284–304, 441–465 [*PRPE, II*: 387–428].#

141. Sur l'application du calcul des probabilités à la Statistique. *Bullettin de l'Institut International de Statistique*, 1909, *XVIII*, 220–253 (in English On the application of the calculus of probabilities to statistics, 505–536).#

142. Probability. In: *Encyclopaedia Britannica* (XIth ed.), 1911, *XXII*, 376–403.[30]#§

143. Contribution to theory of railway rates, *EJ*, 1911, *21*, 346–370, 551–571 [*PRPE, The laws of increasing and diminishing returns*, *I*, 61–99].[31]

144. Monopoly and differential prices. *EJ*, 1911, *21*, 143–148 [*PRPE, Use of differential prices in a regime of competition*, *I*, 100–107]

145. Contribution to theory of railway rates. *EJ*, 1912, *22*, 198–218 [*PRPE, Railway rates*, *I*, 172–191]

146. Contribution to theory of railway rates. *EJ*, 1913, *23*, 206–226 [*PRPE, On some theories due to Professor Pigou*, *II*: 429–449].

147. Professor Moore's *Laws of Wages*. *EJ*, 1912, *22*, 317–323.

148. On the use of the theory of probabilities in statistics relating to society. *JRSS*, 1913, *LXXVI*, 165–193.§

149. A variant proof of the distribution of velocities in a molecular chaos. *PHM*, 1913, *XXV*, 106–109.§

150. A method of representing statistics by analytical geometry. In: *Proceedings of the Fifth International Congress of Mathematicians*, 1913, *II*, 427–440.

151. On the use of analytical geometry to represent certain kinds of statistics. *JRSS*, 1914, *LXXVII*, 300–312; 415–432; 653–671; 724–749; 838–852.#

152. Recent contribution to mathematical economics. *EJ*, 1915, *25*, 36–63, 189–203 [*PRPE, On some theories due to Pareto, Zawadski, W. E. Johnson and others, II*, 450–491].[32]

153. *The Cost of War and Ways of Reducing it Suggested by Economic Theory. A Lecture*, London, Oxford University Press; Humphrey Milford, 1915 (pp. 48).

154. *On the Relations of Political Economy to War: a Lecture*, London, Oxford University Press, Humphrey Milford, 1915 (p. 36).

155. Le relazioni della economia politica con la guerra, *La Riforma Sociale*, 1915, XXVI: 793–820.[33]

156. British incomes and property (Review article of Stamps *The Application of Official Statistics to Economic Problems*). *EJ*, 1916, *26*, 328–336 [*PRPE, III*, 204–212].[34]

157. On the mathematical representation of statistical data. *JRSS*, 1916, *LXXIX*, 455–500.#

158. On the mathematical representation of statistical data. *JRSS*, 1917, *LXXX*, 65–83, 266–288, 411–437.#

159. Some German Economic Writings about the War. *EJ*, 1917, *27*, 238–250 [*PRPE, III*, 215–228].[35]

160. *Currency and Finance in Time of War. A Lecture*, Oxford, Clarendon Press, 1917 (p. 48).

161. After war problems. *EJ*, 1917, *27*, 402–410 [*PRPE, III*, 228–237].

162. Mathematical representation of statistics. A reply. *JRSS*, 1918, *LXXXI*, 322–333.#

163. On the value of a mean as calculated from a sample. *JRSS*, 1918, *LXXXI*, 624–632.#

164. An astronomer on the law of error. *PHM*, 1918, *XXXV*, 422–431.#

165. The doctrine of index numbers according to Professor Wesley Mitchell. *EJ*, 1918, *28*, 176–197 [*PRPE, Professor Wesley Mitchell on index number, I*, 385–405].

166. Psychical research and statistical method. *JRSS*, 1919, *LXXXII*, 222–228.#

167. Discussion on Dr. Bowley's paper. *JRSS*, 1919, *LXXXII*, 365–368.©

168. Methods of graduating taxes on income and capital. *EJ*, 1919, *29*, 138–153 [*PRPE, Graduation of taxes, II*, 243–259].

169. *A Levy on Capital for the Discharge of Debt*. Oxford, Clarendon Press, 1919 (p. 32).

170. La leva sul capitale per il riscatto del debito pubblico. *La Riforma Sociale*, 1919, *30*, 228–251.[36]

171. Entomological statistics. *Metron*, 1920, *I*, 75–82.

172. Mathematical formulae and the Royal Commission on income tax. *EJ*, 1920, *30*, 398–408 [*PRPE, Formulae for graduating taxation, II*, 260–270].

173. Answers to question put by the Royal Commission on Taxation, Royal Commission on Income Tax, 1920, 288, *4*, 581–587, paragraph 11.
174. On the application of probabilities to the movement of gas molecules. *PHM*, 1920, *XL*, 249–272.
175. La teoria pura dell'imposta. In: *Biblioteca dell'economista*, translation by Cesare Jarach, s.n., *XVI*, Torino, *UTET*, 1920, 281–349.[37]@
176. Molecular statistics. *JRSS*, *LXXXIV*, 1921, 71–89.§
177. The genesis of the law of error. *PHM*, 1921, *XLI*, 148–158.#
178. Obituary notice of J. C. Stuart. *EJ*, 1921, *31*, 414–415.
179. On the application of probabilities to the movement of gas-molecules. *PHM*, 1922, *XLIII*, 241–258.©
180. Equal pay to men and women for equal work. *EJ*, 1922, *32*, 431–457.
181. The mathematical economics of Professor Amoroso. *EJ*, 1922, *32*, 400–407.[38]
182. Molecular statistics. *JRSS*, *LXXXV*, 1922, 479–487.
183. The philosophy of chance. *Mind*, 1922, *XXXI*, 257–283.#§
184. Equal pay to men and women for equal work. Presidential address to Section F of the British Association for the Advancement of Science. *BAR*, 1922, 106–132.
185. *Sui metodi per accertare e misurare le variazioni del valore della moneta*, in *Biblioteca dell'economista*, translation by Paolo Conte, s. V, v. XIII, tomo XX, Torino, Utet, 1922, 153–340.[39]@
186. Statistics of examinations. *JRSS*, 1923, *LXXXVI*, 59–60.#
187. Index number according to Mr. Walsh. *EJ*, 1923, *33*, 343–351.
188. Women's wages in relation to economic welfare. *EJ*, 1923, *33*, 487–495.
189. Mr. Correa Walsh on the calculation of index numbers. *JRSS*, 1923, *LXXXVI*, 570–590.
190. On the use of medians for reducing observations relating to several quantities. *PHM*, 1923, *XLVI*, 1074–1088.#
191. Women's wages in relation to economic welfare (Précis of a paper given before Section F). *BAR*, 1923, 461–462.
192. Discussion on Mr. Crump's paper. *JRSS*, 1924, *LXXXVII*, 207–208.©
193. Untried methods of representing frequency. *JRSS*, 1924, *LXXXVII*, 571–594.#
194. The element of probability in index numbers. *JRSS*, 1925, *LXXXVIII*, 557–575.§
195. The revised doctrine of marginal social product. *EJ*, 1925, *35*, 30–39.©
196. The plurality of index numbers. *EJ*, 1925, *35*, 379–388.
197. Reminiscences. In: A. C. Pigou (Ed.), *Memorials of Alfred Marshall* (pp. 66–73). London, MacMillan and Co., 1925.©

198. Mr. Rhodes' curve and the method of adjustment. *JRSS*, 1926, *LXXXIX*, 129–143.#

Section C. Book reviews

1. Jevons, W. S., *The State in Relation to Labour*, AC, July 29 1882, 532, 79.@
2. Mallock, W. H., *Social Equality: a Short Study in a Missing Science*, AC, September 16 1882, 541: 196–197.@
3. Barlow, J. W., *The Ultimatum of Pessimism: an Ethical Study*, AC, March 17 1883, 567: 182–183.@
4. Jevons, W. S., *Methods of Social Reform, and Other Papers*, AC, June 30 1883, 582: 449.@
5. *Studies in Logic* (by members of the Johns Hopkins University Press), AC, October 13 1883, 597: 241–242.@
6. Sidgwick, H., *Fallacies: a View of Logic from the Practical Side*, AC, February 2 1884, 613: 75.@
7. Bradley, F. H., *The Principles of Logic*, AC, May 3 1884, 626: 309.@
8. Keynes, J. N., *Studies and Exercises in Formal Logic*, AC, June 14 1884, 632: 417.@
9. Jevons, W. S., *Investigations in Currency and Finance*, AC, June 19 1884, 637: 38–39.©§
10. Bosanquet, B. (Ed.), *Lotze's System of Philosophy*, AC, February 7 1885, 666: 100–101.@
11. Davidson, W. L., *The Logic of Definition*, AC, October 3 1885, 700: 217.©
12. Sidgwick, H., *The Scope and Method of Economic Science*, AC, May 22 1886, 733: 177.
13. Jevons, W. S., *Journal and Letters of W. Stanley Jevons Edited by his Wife*, AC, May 22 1886, 733: 355–356.
14. Graham, W., *The Social Problem in its Economical, Moral and Political Aspects*, AC, June 24 1886, 742: 50–51.@
15. Whitworth, W. A., *Choice and Chance, 4th edition*, AC, October 16 1886, 754: 254.
16. Bastable, C. F., *The Theory of International Trade*, AC, May 21 1887, 785: 356.
17. Horton, S. D., *The Silver Pound*, AC, July 30 1887, 795: 64–65.©
18. Sargant, W. L., *Inductive Political Economy, vol. 1*, AC, November 5 1887, 809: 298–299.
19. Price, L.L. F. R., *Industrial Peace*, with a Preface by Prof. A. Marshall, AC, December 17 1887, 815: 402–403.©
20. Hastings Berkeley, R. N., *Wealth and Welfare*, AC, 10 March 1888, 827: 164–165.

21. Nicholson, J. S., *A Treatise on Money and Essays on Present Monetary Problems,* AC, May 19 1888, 837: 339.
22. Venn, J., *The Logic of Chance,* AC, August 18 1888, 850: 106.#©
23. Lunt, E. C., *The Present Condition of Economic Science,* AC, September 29 1888, 856: 201.
24. Thorold Rogers, J. E., *The Economic Interpretation of History. Lectures Delivered in Worcester College Hall, Oxford,* AC, December 22 1888, 33 (868): 395–396.
25. Venn, J., *The Logic of Chance,* JED, May 1 1888: 251.©
26. Wicksteed, P. H., *The Alphabet of Economic Science,* AC, February 2 1889, 874: 71.
27. von Böhm-Bawerk, E., *Kapital und Kapitalzins,* AC, May 4 1889, 887: 301.
28. Venn, J., *The Principles of Empirical or Inductive Logic,* AC, May 18 1889, 874: 71.#©
29. Booth, C. (Ed.), *Life and Labour, vol. I,* AC, May 29 1889, 895: 439–440.
30. Graves, R. P., *Life of Sir William Rowan Hamilton, vol. III,* AC, October 26 1889, 912: 262–263.©
31. Cannan, E., *Elementary Political Economy,* JED, January 1 1889: 29.@
32. Wicksteed, P. H., *Alphabet of Economic Science,* JED, March 1 1889: 149.@
33. J. Venn, J., *The Principles of Empirical or Inductive Logic,* JED, June 1 1889: 294–295.©
34. Galton, F., *Natural Inheritance, Nature,* April 25 1889: 603–604.©#§
35. Walras, L., *Éléments d'Économie Politique Pure, Nature,* September 5 1889: 434–436.©§
36. Bertrand, J., *Calcul des Probabilités,* JED, November 1 1889: 592.©#
37. Bertrand, J., *Calcul des Probabilités, Nature,* November 7 1889: 6–7.@
38. Baker, C. W., *Monopolies and the People,* AC, February 8 1890, 927: 95.©
39. Blyth, E. K., *Life of William Ellis,* AC, February 22 1890, 929: 128–129.@
40. Mummery, A., Hobson, J., *The Phisiology of Industry,* JED, April 1 1890: 194.[40]©
41. Giffen, R., *The Growth of Capital,* AC, June 7 1890, 944: 381.
42. Jevons, W. S., *Pure Logic and Other Minor Works,* AC, August 2 1890, 952: 92–93.©
43. Marshall, A., *Principles of Economics, Nature,* August 14 1890, 362–364.©
44. Marshall, A., *Principles of Economics,* AC, August 30 1890, 956: 165–166.
45. von Böhm-Bawerk, E., *Capital and Interest, Nature,* September 11 1890: 462–463.©@
46. von Böhm-Bawerk, E., *Capital and Interest,* English Translation by W. Smart, AC, October 18 1890, 963: 335–336.

47. Longstaff, G. B., *Studies in Statistics,* AC, April 25, 1891, 990: 389–390.
48. Booth, C. (Ed.), *Labour and Life of the People, vol. II,* AC, September 26 1891, 1012: 257.
49. Price, L. L., *Political Economy in England,* JED, May 1, 1891: 282.@
50. Ricardo, D., *Principles of Political Economy,* JED, October 1, 1891: 525–526.©
51. Ely, R. T., *An Introduction to Political Economy,* JED, December 1, 1891: 678.@
52. *The Economic Journal,* vol.1 n. 3, JED, December 1, 1891: 678.@
53. Palgrave, R. H. I. (Ed.), *Dictionary of Political Economy. Part I: Abatement-Bede,* JED, December 1, 1891: 678.@
54. *Keynes, J. N.,* The Scope and Method of Political Economy, *EJ,* 1891, 1:420–423 [PRPE III: 3–7].
55. Marshall, A., Principles of Economics, EJ, 1891, 1: 611–617 [PRPE III: 7–15].
56. Sidgwick, H., *The Elements of Politics,* EJ, 1891, 1: 781–785 [PRPE III: 15–20].
57. Marshall, A., *Elements of Economics of Industry, Nature,* May 12, 1892: 27–28.©
58. Marshall, A., *Elements of Economics of Industry,* JED, June 1, 1892: 320.@
59. Westergaard, W. H., *Die Grundzüge der Theorie der Statistik, Nature,* September 8 1892: 437–438.@
60. Marshall, A., *Principles of Economics,* JED, December 1, 1891: 671–672.©
61. Jacobs, J., *Studies in Jewish Statistics,* EJ, 1892, 2: 135–138.
62. von Böhm-Bawerk, E., (English trans. by William Smart), *The Positive Theory of Capital,* EJ, 1892, 2: 328–336 [PRPE III: 22–31].
63. von Böhm-Bawerk, E., (tr. inglese di William Smart), *The Positive Theory of Capital,* JED, June 1, 1892: 317–318.@
64. Smart, W., *An Introduction to the Theory of Value on the Lines of Menger, Wieser, and Böhm-Bawerk,* EJ, 1892, 2: 336–337 [PRPE III: 31–32].
65. von Düsing, C., *Das Geschlechtverhältniss der Geburten in Preussen,* EJ, 1892, 2: 337–340 [PRPE III: 32–36].
66. Wilcox, W. F., *The Divorce Problem,* EJ, 1892, 2: 341–342.
67. Benson, M., *Capital, Labour and Trade, and the Outlook,* EJ, 1892, 2:342–344 [PRPE III: 20–22].
68. Martin, J. B., *The Grasshopper in Lombard Street,* EJ, 1892, 2: 344.
69. Palgrave, R. H. I. (Ed.), *Dictionary of Political Economy, parts II and III,* EJ, 1892, 2: 524–525.
70. Cossa, L., *Introduzione allo studio dell'economia politica,* EJ, 1892, 2: 685–687.

71. Cantillon, *Essai sur le commerce*, EJ, 1892, 2: 687.
72. Fisher, I., *Mathematical Investigations in the Theory of Value and Prices*, EJ, 1893, 3: 108–112 [PRPE III: 36–41].
73. Thompson, H. M., *The Theory of Wages and its Applications to the Eight-Hours Question and other Labour Problems*, EJ, 1893, 3: 113–115.
74. Rawson, R., *Analysis of the Maritime Trade of the United Kingdom*, EJ, 1893, 3: 115–116.
75. Smart, W., *Women's Wages*, EJ, 1893, 3: 118–119 [PRPE III: 58–59].
76. Martin, J. B., *The Currency of the United States*, EJ, 1893, 3: 119.
77. Hertska, T., *Freiland, ein soziales Zukunftsbild*, EJ, 1893, 3: 284–285.
78. Haynes, F. E., *The Reciprocity Treaty with Canada of 1854*, EJ, 1893, 3: 286.
79. Walsh, C., *Bimetallism and Monometallis: What They Are, and How They Bear upon the Irish Land Question*, EJ, 1893, 3: 286.
80. Bonar, J., *Philosophy and Political Economy in Some of their Historical Relations, Mind,* 1893, II n.s.: 520–525.§
81. Cossa, L., *Introduction to the Study of Political Economy*, translated by Louis Dyer, JED, February 1, 1894: 101–102.
82. von Wieser, F., *Natural Value*, edited by W. Smart, EJ, 1894, 4: 279–285. [PRPE III: 50–58]
83. von Wieser, F., *Natural Value*, JED, April 4, 1894: 221.@
84. Webb, S., Webb, B., *The History of Trade Unionism, vol. 1.*, EJ, 1894, 4: 497–500. [PRPE III: 47–50]
85. Commons, J. R., *The Distribution of Wealth*, EJ, 1894, 4: 684–687.
86. Spyers, T. G., *The Labour Question*, EJ, 1894, 4: 687–688.
87. Brassey, *Papers and Addresses (Work and Wages)*, edited by J. Potter, EJ, 1894, 4: 688–689.
88. *Bulletin de l'Institut International de Statistique, Tomo VII*, EJ, 1894, 4: 689–690.
89. Westegaard, H., *Ueber den Einfluss der Vererbung auf die Sterblichkeit*, EJ, 1894, 4: 691.
90. Duckworth, A., *A Comparison of Populations and Rates of Mortality in New South Wales and Victori; in Sydney and Melbourne*, EJ, 1894, 4: 691–692.
91. Korösi, J., Thirring, G., *Die Haupstadt Budapest im Jahre 1891*, EJ, 1894, 4: 692.
92. Ashley, W.T (Ed.), *Economic Classics. Adam Smith, Ricardo, Malthus*, EJ, 1895, 5: 257–258.
93. Mayr, G. V., *Statistik und GesellSchaftslehre*, EJ, 1895, 5: 258–259.
94. Nitti, F. S., *La misura delle variazioni di valore della moneta*, EJ, 1895, 5: 259–260.

95. Maclean, S. J., *The Tariff History of Canada*, EJ, 1895, 5: 260–261.
96. Atkinson, E., *The Use and Abuse of Legal Tender Acts. True and false Bimetallism*, EJ, 1895, 5: 261–262.
97. Jamieson, G., *The Silver Question. Injury to British Trade and Manufacturers*, EJ, 1895, 5: 395–396.
98. Marshall, A., *Principles of Economics, Third Edition*, EJ, 1895, 5: 585–589 [PRPE III: 64–69].
99. Price, L. L., *Money and its Relations to Prices*, EJ, 1896, 6: 242–245 [PRPE III: 69–73].
100. Nicholson, J. S., *Strikes and Social Problems*, EJ, 1896, 6: 245–248 [PRPE III: 73–77].
101. Pierson, N. G., *Leerbock der Staathuishoudkunde (Eerste Decl.)*, EJ, 1896, 6: 435–437 [PRPE III: 77–80].
102. Six Oxford Men, *Essays in Liberalism*, EJ, 1897, 7: 257.
103. Bastable, C. F., *The Theory of International Trade, with some of its Applications to Economic Policy*, EJ, 1897, 7: 397–403.
104. Graziani, A., *Istituzioni di scienza delle finanze*, EJ, 1897, 7: 403–408 [PRPE III: 80–85].
105. Pierson, N. G., *Leerbock der Staathuishoudkunde Tweede Decl.*, EJ, 1897, 7: 579–582 [PRPE III: 85–90].
106. Darwin, L., *Bimetallism. A Summary and Examination of the Arguments for and against a Bimetallic System of Currency*, EJ, 1898, 8: 105–111.
107. Cournot, A. A., *Researches into the Mathematical Principles of the Theory of Wealth*, English translation by di N. T. Bacon with a bibliography of I. Fisher; Fisher, I., *A Brief Introduction to the Infinitesimal Calculus, designed especially to aid in reading Mathematical Economics and Statistics*, EJ, 1898, 8: 111–114 [PRPE III: 109–112].
108. de Cérenville, M., *Les Impôts en Suisse*, EJ, 1898, 8: 524–525.
109. Davidson, J., *The Bargain Theory of Wages*, EJ, 1899, 9: 229–232.
110. MacFarlane, C. W., *Value and Distribution: an Historical, Critical and Constructive Study in Economic Theory*, EJ, 1899, 9: 233–236.
111. A German Coal Miner, *How the English Workman Lives*, EJ, 1899, 9: 440–441.
112. Vidaurre y Orueta, C., *Economia Politica*, EJ, 1899, 9: 441–442.
113. Ferrin Weber, A., *The Growth of Cities in the Nineteenth century. A Study in Statistics*, EJ, 1899, 9: 558–560.
114. Virgilii, F., Garibaldi, C., *Introduzione alla economia matematica*, EJ, 1899, 9: 560.
115. Bonar, J., Hollander, J. H. (eds.), *Letters of David Ricardo to Hutches Trower and Others, 1811–1823*, EJ, 1900, 10: 221–223 [PRPE III: 95–97].

116. Bastable, C. F., *The Theory of International Trade, with some of its Applications to Economic Policy, 3rd edition*, EJ, 1900, 10: 389–393.

117. Clark, J. B., *The distribution of Wealth. A Theory of Wages and Interest*, EJ, 1900, 10: 534–537 [PRPE III: 97–101].

118. Smart, W., *Taxation of Land Values and the Single Tax*, EJ, 1900, 10: 538.

119. Hollander, J. (Ed.) *Studies in State Taxation, with particular reference to the Southern States*, EJ, 1900, 10: 538–539.

120. Gide, C., *La Co-opération (Conférences de propagande)*, EJ, 1902, 12: 89–90 [PRPE III: 112–113].

121. Wells, H. G., *Anticipations of the Reaction of Mechanical and Scientific Progress upon Human Life and Thought*, EJ, 1902, 12: 90–92 [PRPE III: 113–115].

122. Webb, B., Webb, S., *The History of Trade Unionism*; IDEM, *Industrial Democracy*, EJ, 1902, 12: 257–259.

123. Schloss, D., *Les modes de Rémuneration du Travail*, EJ, 1902, 12: 259–260.

124. Cox, H., *The United Kingdom and its Trade*; Chiozza, L. G., *British Trade and the Zollverein Issue*, EJ, 1902, 12: 401–402.

125. Bastable, C. F., *Public Finance*, EJ, 1903, 13: 226–228 [PRPE III: 41–43].

126. Saint-Leon, E. M., *Cartels and Trusts*, EJ, 1903, 13: 228–230.

127. von Bortkiewiez, L., Anwendungen der Wahrscheinlichkeitsrechnung auf Statistik in *Encyklopädie der mathematischen Wisenschaften*; Pareto, V., Anwendungen der Mathematik auf Nationalökonomie in *Encyklopädie der mathematischen Wisenschaften*, EJ, 1903, 13: 230–233. [PRPE III: 43–47]

128. Root, J. W., *The Trade Relations of the British Empire*, EJ, 1903, 13: 375–382.

129. Ashley, W. J., *The Tariff Problem*, EJ, 1903, 13: 571–575.

130. Cannan, E., *A History of the Theories of Production and Distribution in English Political Economy from 1776–1848*, EJ, 1903, 13: 624–627 [PRPE III: 115–119].

131. Pigou, A. C., *The Riddle of the Tariff*, EJ, 1904, 14: 65–67 [PRPE III: 119–121].

132. Nicholson, J. S., *Elements of Political Economy*, EJ, 1904, 14: 67–69 [PRPE III: 121–123].

133. Bowley, A. L., *National Progress in Wealth and Trade*, EJ, 1904, 14: 268–271 [PRPE III: 123–126].

134. Plunkett, H., *Ireland in the New Century*, EJ, 1904, 14: 430–433 [PRPE III: 126–128].

135. Norton, J. P., *The Theory of Loan Credit in Relation to Corporation Economics*, EJ, 1904, 14: 604–605 [PRPE III: 128–130].

136. Graziani, A., *Istituzioni di economia politica*, EJ, 1904, 14: 605–607 [PRPE III: 130–132].
137. Dalla Volta, R., *Sulla ripercussione e la incidenza dei dazi doganali*, EJ, 1904, 14: 608.
138. Dietzel, H., *Vergeltungzölle,* EJ, 1904, 14: 608–609 [PRPE III: 132–133].
139. Nicholson, J. S., *The History of the English Corn Laws*, EJ, 1905, 15: 60–62 [PRPE III: 133–136].
140. Cunynghame, H., *A Geometrical Political Economy. Being an Elementary Treatise on the Method of Explaining Some of the Theories of Pure Economic Science by Means of Diagrams*, EJ, 1905, 15: 62–71 [PRPE III: 136–144].
141. Carver, T. N., *The Theory of Distribution*, EJ, 1905, 15: 71–72 [PRPE III: 144–145].
142. Taussig, F. W., *The Present Position of the Doctrine of Free Trade*, EJ, 1905, 15: 208–211 [PRPE III: 145–147].
143. E. M. S., *Henry Sidgwick. A Memoir*, EJ, 1906, 16: 273–278 [PRPE III: 147–152].
144. Amery, L. S., *The Fundamental Fallacies of Free Trade*, EJ, 1906, 16: 568–573.
145. Cohen, J., *Chance: A Comparison of Facts with the Theory of Probabilities*, JRSS, 1906: 227–228.#©
146. de Foville, A., *La Monnaie*; Guyot, Y., *La Science Economique*, EJ, 1907, 17: 407–412 [PRPE III: 152–157].
147. Andréadès, A., *PERI APOGRAFHS*, EJ, 1908, 18: 308–309 [PRPE III: 157–158].
148. Mitchell, W. C., *Gold Prices and Wages under the Greenback Standard,* EJ, 1908, 18: 578–582 [PRPE III: 158–162].
149. Dietzel, H., *Bedeutet Export von Produktionsmitteln Volkswirtschaftlichen Selbstmord? Unter besonder Berücksichtigung des Maschinen und Kohlenexports Englands*, EJ, 1909, 19: 100–102.
150. Rea, R., *Free Trade in Being*, EJ, 1909, 19: 102–106 [PRPE III: 164–168].
151. Withers, H., *The Meaning of Money*, EJ, 1909, 19: 251–253 [PRPE III: 162–164].
152. Jevons, W. S., *Investigations in Currency and Finance*, EJ, 1909, 19: 253.
153. Benini, R., *Principi di statistica metodologica*, JRSS 1909: 104–110.#§©
154. Colson, C., *Cours d'Économie Politique professé à l'École Nationale des Ponts et Chaussés*, EJ, 1910, 20: 57–63 [PRPE III: 168–175].
155. Mill, J. S., *Principles of Political Economy*, a cura di W. J. Ashley, EJ, 1910, 20: 101–102 [PRPE III: 175–176].

156. Clark, J. M., *Standards of Reasonableness in Local Freight Discriminations*, EJ, 1910, 20: 604–606 [PRPE III: 176–179].

157. Hammond, M. B., *Railway Rate Theories of the Interstate Commerce Commission*, EJ, 1911, 21: 601–603 [PRPE III: 179–181].

158. Moore, H. L., *Laws of Wages. An Essay in Statistical Economics*, EJ, 1912, 22: 66–71.§

159. Levy, H., *Monopoly and Competition: a Study in English Industrial Organization*, EJ, 1912, 22: 89–90.

160. Pigou, A. C., *Wealth and Welfare*, EJ, 1913, 23: 62–70 [PRPE III: 181–189].

161. Zawadski, W., *Les Mathématiques appliquées à l'Economie Politique*, JRSS 1914: 754–757.#©

162. Alberti, M., *L'economia del mondo prima e dopo la guerra europea*, EJ, 1915, 25: 395–398.

163. Economists on war: Sombart, W., *Handler und Helden*; Nicholson, J. S., *The Neutrality of the United States in Relation to the British and German Empires*; Seligman, E. R. A., *An Economic Interpretation of the War*; Guyot, Y., *Les Causes et les Consequences de la Guerre*, EJ, 1915, 25: 604–610 [PRPE III: 194–201].

164. Pigou, A. C., *The Economy and Finance of the War*, EJ, 1916, 26: 223–227 [PRPE III: 189–194].

165. Preziosi, G., *La Germania alla Conquista dell'Italia*, EJ, 1916, 26: 230–233 [PRPE III: 201–204].

166. Gill, C., *National Power and Prosperity: A Study of the Economic Causes of Modern Warfare*, EJ, 1917, 27: 96–98 [PRPE III: 212–214].

167. Lehfeldt, R. A., *Economics in the Light of War*, EJ, 1917, 27: 98–99.

168. *War Finance Primer*, EJ, 1917, 27: 400–401.

169. Seligman, E. R. A., *A Constructive Criticism of the United States War Tax Bill*, EJ, 1917, 27: 401.

170. Westergaard, H., *Scope and Methods of Statistics*, JRSS, 1917: 546–551.#©

171. Anderson, B. M., *The Value of Money*, EJ, 1918, 28: 66–69 [PRPE III: 237–241].

172. Moulton, H. G., *Principles of Money and Banking;* Phillips, C. A., *Readings in Money and Banking*, EJ, 1918, 28: 70–72 [PRPE III: 241–243].

173. Smith-Gordon, L., Staples, L. C., *Rural Reconstruction in Ireland: a Record of Cooperative Organisation*; A. E. (George Russell), *The National Being*, EJ, 1918, 28: 198–202 [PRPE III: 243–248].

174. Loria, A., *The Economic Causes of The War*, EJ, 1918, 28: 317–320 [PRPE III: 260–262].

175. Arias, G., *Principi di Economia Commerciale*, EJ, 1918, 28: 327–330 [PRPE III: 263–266].
176. Bernis, F., *La Hacienda Española. Los Impuestos*, EJ, 1919, 29: 83–85.
177. Cannan, E., *Money, Its Connexion with Rising and Falling Prices*, EJ, 1919, 29: 214–216 [PRPE III: 248–251].
178. Andréadès, A., *ΣΨΣΤΗΜΑ ΔΗΜΟΣΙΑΣ ΟΙΚΟΝΟΜΙΚΗΣ ΤΟΜΟΣ Α. ΜΕΡΟΣ Β. ΙΣΤΟΡΙΑ ΤΗΣ ΕΛΛΗΝΙΚΗΣ ΔΗΜΟΣΙΑΣ ΟΙΚΟΝΟΜΙΑΣ ΑΠΟ ΤΩΝ ΗΠΩΙΚΟΝ ΧΡΟΝΩΝ ΜΕΧΡΙ ΤΗΣ ΣΨΣΤΑΣΕΩΣ ΤΟΨ ΕΛΛΗΝΙΚΟΨ ΒΑΣΙΛΕΙΟΨ. ΨΠΟ ΑΝΔΠΕΟΨ ΜΙΧ. ΑΝΔΠΕΑΔΟΨ* (The economic institutions of ancient Greece), EJ, 1919, 29: 217–219 [PRPE III: 251–253].
179. Lehfeldt, R. A., *Gold Prices and the Witwatersrand*, EJ, 1919, 29: 327–330 [PRPE III: 253–257].
180. *Interim Report of the National Industrial Conference*, EJ 1920, 29: 478–480.
181. Webb, S., Webb. B., *The History of Trade Unionism*, EJ, 1920, 30: 219–222 [PRPE III: 257–260].
182. Cassel, G., *Theoretische Sozialökonomie*, EJ, 1920, 30: 530–536 [PRPE III: 266–272].
183. Bowley, A. L., *The Change in Distribution of the National Income, 1880–1913*, JRSS 1920: 482–484.§©
184. Nicholson, J. S., *The Revival of Marxism*; Loria, A., *Karl Marx*, EJ, 1921, 31: 71–73 [PRPE III: 273–275].
185. Hoare, A., *The National Needs of Britain*, EJ, 1921, 31: 91.
186. Gough, G. W., *Wealth and Work*, EJ, 1921, 31: 380–381.
187. Pigou, A. C., *The Political Economy of War*, EJ, 1922, 32: 73–77.
188. McDougall, W., *National Welfare and National Decay*, EJ, 1922, 32: 84–86.
189. Brand, R. H., *War and National Finance*, EJ, 1922, 32: 217–219.
190. Subercaseaux, G., *Le paier monnaie*; IDEM, *El sistema monetario i la organización bancaria de Chile*, EJ, 1922, 32: 529–533.
191. Keynes, J. M., *A Treatise on Probability*, JRSS, 1922: 107–113.#§©
192. Rathenau, W., *In Days to Come*, traduzione di Eden e Cedar Paul, EJ, 1923, 33: 70–73.
193. Marshall, A., *Money Credit and Commerce*, EJ, 1923, 33: 198–204.
194. Seven members of the Labour Party, *The Labour Party's Aim: A Criticism and a Restatement*, EJ, 1923, 33: 539–542.
195. Bernis, F., *Consequencias economics de la guerra*, EJ, 1923, 33: 544–545.
196. Tschuprow, A. A., *Das Gesetz der grossen Zahlen und der Stochastisch-Statistische Standpunkt in der modernen Wissenschaft*, JRSS, 1923: 65–66.©

197. Bowley, A. L., *The Mathematical Groundwork of Economics*, EJ, 1924, 34: 430–434.
198. Fisher, H. A. L., *The Economic Position of Married Women*, EJ, 1924, 34: 446–449.
199. Czuber, E., *Die Statistischen Forschungsmethoden*, JRSS, 1924: 103–106.#©§
200. Ricardo, D., *Economic Essays*, edited by E. C. K. Gonner, JRSS, 1924: 110–112.©
201. Pigou, A. C., *The Economics of Welfare*, EJ, 1925, 35: 30–39.
202. Clark, J. M., *Studies in the Economics of Overhead Costs*, EJ, 1925, 35: 245–251.
203. Rubio, I., *Matematica de la mortalidad con elementos de probabilades*; de Miguel, A., *Introducion a la metodologia estadistica (Fundamentos de Estad'stica Matemàtica)*, JRSS, 1925, LXXXVIII: 109–110.#©
204. Yule, G. U., *A Mathematical Theory of Evolution based on the Conclusions of Dr. J. C. Willis, F.R.S.*, JRSS, 1925, LXXXVIII: 433–436.©§

Section D. Palgrave's entries

The editorial history of the *Dictionary* is complex. After the first edition – 1894–1896 –, several reprints with corrections appeared until the new one edited by H. Higgs between 1923 and 1926, entitled *Palgrave's Dictionary of Political Economy*. In this new edition the most important corrections were related to the introduction of an appendix with new entries and expanded versions of some old entries.

The volumes we have considered refer to the first edition and the appendix of 1926. To give an idea of the length and the importance of Edgeworth's entries, we have utilized the method adopted in the analytical index of the first edition of the same *Dictionary* (Vol. III, p. 692). Every page has been ideally divided in four quadrants labelled with the letter A–D counterclockwise.

Every entry has been indicated with the page numbers and the quadrants; a quadrant is occupied by an entry if more than five rows of the entry belong to the quadrant. For example: Census 238A–243C indicates that the entry extends from the quadrant A of page 238 to the quadrant C of page 243; Campanella, Tommaso 208 BC indicates that the entry extends in the quadrants B and C of page 208. We indicate with * entries with only a very short definition, or biographical data and bibliography; with ** entries only partially signed by Edgeworth. In the last column we indicate the entries reprinted in the current edition of the *Palgrave's* (EMN, 1987).

Volume I, letters A–E, 1894 (reprints with corrections 1901; 1909; with corrections 1915; 1919; new edition 1925; 1926)

Absentee	3A–4D		EMN: Vol. 1: 2–3
Agents of production	21B–22A		EMN: Vol. 1: 40–41
Aickin, Rev. Joseph	30B	*	
Aleatory	31AB		
Antoninus, St.	43A	*	
Attwood, Thomas and the Birmingham School	67B–68A		reprinted as "Birmingham School" in EMN: 1987, Vol. 1: 248–249
Auxiron, Claude Francois Joseph d'	74A	*	
Average	74AD		
Bayley, Samuel on Value	82C–83A		
Barter and Exchange	122CD		EMN: Vol. 1: 198–199
Bastiat as a Theorist	124BC		
Baxter, Robert Dudley	126BD		
Beldam	129A	*	
Berkeley, George, Bishop of Cloyne	134D–135B		
Birth-rate	150B–151A	©	
Blake, William F.R.S.	153C	*	
Bounties, Abstract theory of	172C–173B		
Brassey, Thomas	176D		
Bright, John	179B		
Brindley, James	179BC	* ©	
Briscor, John	179CD	©	
Brougham, Henry	181BC		EMN: Vol. 1: 279–280
Buckle, Thomas Henry	184D–185B		EMN: Vol. 1: 283
Buquoy, Georg Franzm Count	192D	*	
Buridan, Jean	194D	*	
Burke, Edmund	194D–195C		
Burlamaqui, Jean Jacques	195C		
By-products, Theory of value of	197B–D		
Camerarius, Joachim	208B	*	
Campanella, Tommaso	208B–C		
Canard, Nicholas Francois	209D		
Cantillon, Richard	215D	**	
Cary, John	230AB		

Efficiency of Money	685CD	
Elasticity	691BC	
Eliot, Francis Perceva	692CD	
Ellis, William	693D–694A	
Error, Law of	751B–753D	
Exchange, value in	759D–762C	
Expectation of Life	790B	

Volume II, letters F–M, 1896 (reprint 1896; 1900; with corrections 1906; with corrections 1910; with corrections 1912; 1917; new edition 1923; 1925).

Facts	11C–12A	
Fallacies	17B–18A	©
Fixed Incomes	88D	
Forced Currency	96C	*
Fullarton, John	167BD	
Functions	167D–169A	
Gossen, Hermann Heinrich	231A–233D	
Growth, Proportionate	268BC	
Hagen, Karl Heinrich	272AC	
Hearn, William Edward	294C–295A	
Helferich, Johannes a Renatus von	298BC	
Higgling	304D–305B	EMN: Vol. 2: 652–653
Ideal money	353AC	
Income	374B–375A	
Inconvertible currency	380AC	
Index numbers	384D–387D	
Indifference, Law of	387D–388A	EMN: Vol. 2: 786–787
Intrinsic value	455D–456B	
Jenkin, Henry Charles Fleeming	473AD	
Jennings, Richard	473D–474C	
Jones, Richard	490A–491B	
King, Peter, Lord	506BC	
Least squares, method of	587CD	
Luck	648BD	©
Luxury	653B–655A	
Margin	690B–691A	
Marriage rate	701A–702A	
Mathematical Method in Political Economy	711A–713A	EMN: Vol. 3: 404–405

Maximum satisfaction	717AC		EMN: Vol. 3: 409
Mean	718C	*	
Means, method of	718C	*	
Mill, James	755A–756B		
Mill, John Stuart	756B–763B		
Moffat, Robert Scott	779B	*	
Monopoly	805B–807A		
Multiplication of Services	828BC	*	

Volume III, letters N–Z, 1899 (reprint 1901; with corrections and general appendix A-Z 1908; with corrections 1910; with corrections 1913; 1918; new edition 1926).

Negative quantities	14D–15A		EMN: Vol. 3: 624
Numerical determination of the laws of utility	28AC		EMN: Vol. 3: 687
Over-production	45C–46A		
Peacock, George	82D–83A		
Playfair, William	116C–117B	**	EMN: Vol. 3: 895–896
Pleasure and pain	117BD		EMN: Vol. 3: 896
Porphyry	170BC	*	
Present goods	187CD		
Probability and calculus of probability	208AC	©	
Rae, John	250A–251A		
Risk	314BC		
Supply curves	497D–498A	©	
Total Utility	551D–552A		
Unit of Value	599D–600A		
Utility	602AB		
Wealth	660B–661B		
Wilson, Glocester	669BC	*	

Appendix Vol. I (new edition of 1925)

Averages	818D	
Birth-rate	823C	
Census	834BD	
Curves	861C	
Economics, Teaching at Oxford	879D–881D	Present also in the appendix in the third volume of 1908 edition at the pages 729B–731A.

Error, Law of   886BC

Appendix Vol. II (new edition of 1923)

Index numbers   895B–896D

Appendix Vol. III (new edition of 1926)

Pantaleoni, Maffeo   709C–710D
Pareto, Vilfredo   710D–711D
Pareto's law   712A–713B

NOTES

[*] Researches have been conducted for the most part in Italy, at the Biblioteca Nazionale Centrale of Florence, where I have consulted the reviews collected in the Fondo Benn; I have consulted also some reviews at the Biblioteca Nazionale Centrale of Rome and in the Library of the Fondazione Luigi Einaudi of Turin. To complete some items I have been in the Biblioteque Nationale de France, thanks to a grant of the Consiglio Nazionale delle Ricerche (CNR). Michelangelo Vasta has allowed me to have access to the sources of the Bodleian Library of Oxford.

My bibliographical work is built on the work of the other scholars, extending their coverage and correcting some errors, but making great use of their invaluable efforts. My principal point of reference has been Johnson's bibliography (1954) that I have been able to have – thanks to Peter Newman – with some annotations by Stephen Stigler. I would like to thank Marco Dardi, Peter Newman, Warren J. Samuels, Stephen Stigler and two anonymous referees. I am the sole person responsible, however, for any remaining errors or omissions.

1. A selective list of writings on Edgeworth must comprehend at least Bowley (1928), Keynes (1936); Stigler (1978, 1986, 1999); Creedy (1986); Newman (1987); but see also Blaug (1992). The texts of Newman and Stigler offer a very important guide to the reading of Edgeworth's original papers in economics and statistics respectively. For Edgeworth's theory of probability and expectation, see Baccini (1997, 2001); for Edgeworth's reviews, see Newman (1990).

2. McCann's volumes contain the photographic reprints of a total of 89 Edgeworth's papers and reviews. For a critical point of view about the volume see Newman, 1997. Mirowski's book contains the re-edition of 22 articles, 12 book reviews, *Metretike*, and selected unpublished correspondence with Henry Ludwell Moore and Edwin Bidwell Wilson. It has been reviewed by Stigler (1995). With the only exception of *Metretike,* all the papers included have been edited down, in some cases, drastically. The original pagination is not included, so in order to be able to follow the internal references – Edgeworth was a great one for self-citation (Stigler 1995, p. 803) – we must read the original papers.

3. Mirowski has attributed to Edgeworth some book reviews published in *Academy* between 1888 and 1889. None of Mirowski's attributions has been included in our bibliography. For a discussion, see the Appendix to this paper.

4. I have inserted, and counted, in section A and B the translations of Edgeworth's articles published during his life. I have not inserted the, already cited, selected

correspondance published in Mirowski (1994, pp. 427–439); the Syllabus for 1885 lectures, King's College, London, entitled The logic of statistics, edited by Stigler in Stigler (1986, pp. 363–366); the Syllabus for 1892 Newmarch lectures, University College, London, entitled On the uses and methods of statistics, edited by Stigler in Stigler (1986, pp. 367–369). Stephen Stigler signalled that there are 8 Edgeworth's bets published in *The Text of the Second Betting Book of All Souls College*, with a preface, commentary and elucidation by Charles Oman, Oxford, Privately printed for members of the College only, 1938. Edgeworth's bets – not inserted in the bibliography- are at the number 418, 463, 493, 564, 568, 582, 589, 738, 748.

5. Only one item inserted by Johnson amongst articles has been listed here in the book reviews section (review n. 80).

6. The papers anthologized in PRPE are selected, edited and revised (Bowley, 1928, p. 1) by Edgeworth himself. Newman (1997, p. 353) wrote: In some ways it was unfortunate that [the Royal Economic Society] asked to [Edgeworth] to edit the collection, since he omitted some important articles and many reviews, and quite severely edited (and in one case altered) some of the articles that did appear. So in every case Edgeworth's students must refer also to the original papers.

7. Italian translation by Valentino Dominedò, *Psichica matematica, Saggio sull'applicazione delle matematiche alle scienze morali*, in Del Vecchio, G. (Ed.), *Economia Pura*, Torino, UTET, 1937, pp. 191–327.

8. J. W. L. Glaisher signed a summary of Edgeworth's paper on the *Proceedings of the Cambridge Philosophical Society,* 1885, 5, pp. 310–312. Johnson ascribed this abstract to Edgeworth.

9. This paper is the abstract of book 3.

10. This paper is the reply to Venn, J., Letter to the editor: the law of error, *Nature*, 1887, pp. 411–412.

11. This paper is the abstract of 47.

12 This paper is the same as numbers 53 and 54.

13. This paper is the same as numbers 52 and 54.

14. This paper is the same as numbers 52 and 53.

15. This paper contains the conclusions of 51.

16. This is the reply to the paper by Arthur Berry, Alcune brevi parole sulla teoria del baratto di A. Marshall, GE, 1891, I, pp. 549–553.

17. Johnson dated the paper in 1895. This is the reply to Bortkewitsch, L. V., Kritische Betrachtungen zur theoretischen Statistik, «Jahrbücher für Nationalökonomie und Statistick», 1895, X, pp. 321–360.

18. This is the very short summary of 90 and 91.

19. The English reprint in PRPE is the translation from Italian version "because the original English version was lost" (PRPE I: 111).

20. This paper is the summary of 104, 107 and 112.

21. Newman (1990, p. 131) counted this article as a review.

22. These [105, 107] (without the Supplement [112]) were issued in a volume for private circulation together with xix pages of Introductory Description and 1 page of Corrigenda and Addenda (Bowley, 1928, p. 135).

23. Newman (1990, p. 131) counted this article as a review.

24. This entry has been inserted lately in 142.

25. Circulated privately with unpublished Continuation of appendix (Bowley, 1928: 135; Kendall, Doig, 1968).

26. Signalled by Stephen Stigler.

27. This text, published for the first time in PRPE, according to Edgeworth (PRPE, II, p. 227) "[reproduces] the substance of an Address given to the Students' Union at the London School of Economics, January, 1906".

28. This paper is the summary of 132 and 133.

29. Italian translation of 132 and 133.

30. As noted in the text and in the final note of the entry (p. 403), the paragraphs §§41, 52, 62, 69, 72 and 76–93, reproduced the paragraphs of the IXth edition by Morgan Crofton.

31. Italian translation of the papers 143, 145, 146 by M. Del Vescovo, *Contributi alla teoria delle tariffe ferroviarie*, Roma, Università degli Studi «La Sapienza», Dipartimento di teoria economica e metodi quantitativi per le scelte pubbliche, 1992.

32. Newman (1990, p. 131) counted this article as a review.

33. Italian version of 154; published also as pamphlet, Torino, Società Tipografica Editrice Nazionale, 1919.

34. Newman (1990, p. 131) counted this article as a review.

35. Newman (1990, p. 131) counted this article as a review.

36. Italian version of 169.

37. Italian version of 100.

38. Newman (1990, p. 131) counted this article as a review.

39. Italian version of 18, 19 and 51.

40. See Newman (1987, p. 89).

REFERENCES

Baccini, A. (1997). Edgeworth on the fundamentals of choice under uncertainty. *History of Economic Ideas*, *2*, 27–51.

Baccini, A. (2001). Frequentist probability and choice under uncertainty. *History of Political Economy*, *33*(4), 743–772.

Blaug, M. (Ed.) (1992). *Alfred Marshall and Francis Edgeworth*. Elgar Reference Collection Series. Pioneers in Economics series, 29, Aldershot, Elgar.

Bowley, A. L. (1928). *F. Y. Edgeworth's Contributions to Mathematical Statistics*. London: Royal Statistical Society. (Clifton, N. J., A. M. Kelley, 1972)

Creedy, J. (1986). *Edgeworth and the Development of Neoclassical Economics*. London: Basil Blackwell.

Eatwell, J., Milgate, M., & Newman, P. (Eds) (1987). *The New Palgrave: A Dictionary of Economics*. London: MacMillan.

Hildreth, C. (1968). Edgeworth, Francis Ysidro. In: *International Encyclopedia of the Social Sciences* (pp. 506–509). MacMillan.

Johnson, H. G. (1954). F. Y. Edgeworth: a bibliography. Chicago, mimeo.

Kendall, M. G., & Doig, A. (1968). *Bibliography of Statistical Literature Pre-1940*. Edinburgh: Oliver & Boyd.

Keynes, J. M. (1926). Obituary: Francis Ysidro Edgeworth, 1845–1926, *Economic Journal*, *36*, 140–153. [*Essays in Biography*, Vol. X of *Collected Writings*, London: Macmillan, 1972, 251–266.]

McCann, C. R. (Ed.) (1996). *F. Y. Edgeworth: Writings in Probability and Statistics* (Vol. 3). Cheltenham: Edward Elgar.

Mirowski, P. (Ed.) (1994). *Edgeworth on Chance, Economic Hazard, and Statistics*. Boston: Rowman & Littlefield.

Newman, P. (1987) Edgeworth, Francis Ysidro in *EMN*, *II*, 84–98.

Newman, P. (1990). Reviews by Edgeworth, in Hey, J. D., Winch, D. (eds.), *A Century of Economics*, London: Basil Blackwell: 109–141.

Newman, P (1997) F. Y. Edgeworth: Writings in Probability, Statistics and Economics. C. R. McCann (Ed.). *The European Journal of the History of Economic Thought*, 4(2), 353–359.

Stigler, S. M. (1978). Francis Ysidro Edgeworth, statistician. *JRSS* (A), *141*, 287–322.

Stigler, S. M. (1986). *The History of Statistics. The Measurement of Uncertainty before 1900*, Cambridge, MA: Belknap Press of Harvard University Press.

Stigler, S. M. (1995). Edgeworth on Chance, Economic Hazard, and Statistics. P. Mirowski (Ed.). *Journal of the American Statistical Association*, *90*, 803–804.

Stigler, S. M. (1999). *Statistics on the Table*. Cambridge Mass. and London: Harvard University Press.

APPENDIX

Mirowski (1994) attributes to Edgeworth some reviews published in *Academy*. The aim of this appendix is to cast doubt on this question, without claiming, however, to arrive at any definitive conclusions regarding Edgeworth's authorship.

In *Academy*, September 1, 1888: 132–133 we find, under the title Some economical books, six unsigned book reviews, and more precisely, the reviews of L. Verney, *How the Peasant Owner Lives in Parts of France, Germany, Italy and Russia*; W. O'Connor Morris, *The Land System of Ireland*; M. Frewen, *The Economic Crisis*; E. C. K. Gonner, *Political Economy. An Elementary Text-book of the Economics of Commerce*; J. B. Clark and F. H. Giddins, *The Modern Distributive Process*; L. Cossa, *Taxation, Its Principles and Methods* translated by H. White. Mirowski attributes to Edgeworth the reviews of Gonner, Clark-Giddins and Cossa.

In *Academy*, March 1, 1890: 149–150, under the heading Two Foreign Books on Economics, we find the unsigned reviews of T. Hertska, *Freiland, ein soziales Zukunftsbild*, and of L. Walras, *Éléments d'Économie Politique*. Mirowski attributes to Edgeworth the review of Walras only. Edgeworth reviewed the first book in 1893 for the EJ, and the second in 1889 for *Nature*.

In *Academy* (April 12, 1890: 251–252), under the title Some Books on Economics, there are six unsigned review of the following books: W. Donisthorpe, *Individualism: a System of Politics*; W. S. Dabney, *The Public Regulation of Railways*; H. D. MacLeod, *The Theory of Credit*; F. Minton, *The Welfare of the Millions*; A. F. Mummery and G. A. Hobson, *The Physiology of Industry*; J. Mavor, *Economic History and Theory*. Mirowski attributes to Edgeworth the reviews of Dabney, MacLeod, Minton, and Mummery and Hobson.

In *Academy,* September 13, 1890: 218–220, under the title Some economical books we find five unsigned reviews of the following books: E. Atkinson, *The Industrial Progress of The Nation,* F. A. Walker, *First Lessons in Political Economy*; F. W. Bain, *Occam's Razor*; F. W. Newman, *Political Economy*; C. J. Daniell, *The Industrial Competition of Asia.* Mirowski attributes to Edgeworth the reviews of Walker, Bain, and Newman.

In 1889 under the same headings, there are also other unsigned reviews which Mirowski does not, however, attribute to Edgeworth. In *Academy,* July 27, 1889: 53–54, the following reviews appear under the title Some foreign books on political economy: R. Auspitz and R. Lieben, *Untersuchungen über die Theorie des Preises;* G. Schmoller, *Zur Literaturgeschichte der Staats und Sozialwissenschaften*; F. Quesnay, *Oeuvres économique et philosophique*, edited by A. Oncken; *Archiv für Soziale Gesetzebung und Statistik.*

In *Academy*, September 7 1889: 148–149, under the title Some books on economics, there are the reviews of the following seven books: W. R. Sorley, *Mining Royalties*; F. A. Minton, *Capital and Wages*; F. W. Taussig, *The Tariff History of the United States*; A. Newsholme, *Vital Statistics*; W. Leighton, *The Standard of Value*; E. J. Donnel, *Outlines of a new Science*; Zucherkandl, *Zur Theorie des Preises.*

For the reviews in *Academy*, Edgeworth's authorship should be proved through reference to archival documents, or through a very detailed comparative textual analysis of the style of the reviews. All the *external* evidence testifies *against* Edgeworth's authorship. From Mirowski's attributions it is possible at least to infer that: (i) in the same years there was at least another economist who reviewed for the *Academy*; (ii) this economist reviewed also foreign books. It is moreover worthwhile to notice that; (iii) in the years 1888–1889, Edgeworth contributed regularly to *Academy* with signed reviews; (iv) almost all of Edgeworth's signed reviews were indexed in the first page of *Academy*; (v) sometimes Edgeworth's review was the first article of the issue; (vi) with only one exception (review n. 40), none of the reviews attributed to Edgeworth by Mirowski refer to books reviewed by Edgeworth elsewhere, while in the same years Edgeworth usually published reviews of the same book in different journals.